WEST AND SMITH'S

LAW OF DILAPIDATIONS

ELEVENTH EDITION

by

PF SMITH, BCL, MA

Reader in Property Law
The University of Reading

With a Foreword by

PE GOODACRE, RD, MSc, FRICS, FCIOB
Principal, College of Estate Management

A division of Reed Business Information
Estates Gazette
151 Wardour Street, London W1F 8BN

Eleventh Edition 2001

ISBN 0 7282 0352 9

© College of Estate Management 2001

Typesetting by Amy Boyle, Rochester, Kent
Printed in Finland by WS Bookwell Ltd

Foreword

For a textbook to be approaching its 70th anniversary is remarkable and says two things.

First, that the original conception and need for the textbook was most appropriate, yet I doubt whether Beniah Adkin, the second Principal of the College of Estate Management, who was the original author, would have anticipated that it would now be approaching its 11th edition.

The second factor is the skill with which successive colleagues of the College have been able to prepare new editions.

This 11th edition quite properly incorporates the name of Peter Smith in the title; Peter, in fact, has been involved with both the 9th and 10th editions. This 11th edition is entirely Peter's authorship and I consider the property profession to be most fortunate in having such a committed leading academic lawyer prepared to devote so much time to this major revision.

The law of the land constantly evolves, meeting changes in policy and practice. That the law should evolve in such a way is important to society in that it can operate in a fair and equitable framework. I commend this book to its readership with no false modesty in describing it as the leading shorter work in the field.

Peter Goodacre
Principal, College of Estate Management
August 2000

Preface

This book will celebrate its 70th anniversary in 2003. Although the law of dilapidations has become more complex and diverse during this period, the fundamental basis of some of the rules would not be unrecognisable when compared to that applying when the first edition appeared. It is the detail, breadth and complexity of the law which has greatly increased. Many of the principles arise out of the contractual relationship of landlord and tenant. The scope of the wide-ranging subject of dilapidations extends, however, to freehold owners. Dilapidations issues indeed affect many matters, such as occupiers' liability, rent review, business tenancy renewals, fixtures, and party walls, to name but some of the areas examined in this book.

The text required substantial revision, not only to bring it up to date but also following on reconsideration of the text as a whole. In connection with updating, there has been a string of recent cases concerned with the interpretation of tenants' repairing obligations and as to service charge recovery. We have also seen significant developments in relation to remedies, notably damages, specific performance and the new entrant of repudiation. The statutory implied obligations of residential landlords continue to develop. We have seen the extension of the legislation applying to Inner London party walls to the whole country, and the removal of the old dual regime as a result.

Reform of the law is under consideration. There is discussed in various places both the Law Commission's important final Report *Responsibility for the State and Condition of Property* (1996), and the government Green Paper "Quality and Choice: A Decent Home for All". During the life of the present edition, we could well see the enactment of additional legislative remedies for dissatisfied long residential lessees against defaulting landlords, as well as of a commonhold scheme to enable the development of freehold flats under a regime of mutual enforceability of repairing and maintenance obligations. A Commonhold and Leasehold Reform Bill received its First Reading in the House of Lords at the end of

December 2000. Structural changes have been made to the text. The material dealing with tenant liability to repair is now in a single chapter. For the first time, selective comparative law material has been included, where it aids the understanding of our principles. As might be expected, a good deal of this material is drawn from Commonwealth jurisdictions. However, particularly in evaluating reform, it seemed appropriate to insert a modest amount of French material, if only because of the fact that the starting point of civilian systems is different to the *laissez-faire* approach of the common law.

I am grateful to Peter Goodacre, Principal, College of Estate Management, for his support and encouragement in connection with this 11th edition and for contributing the foreword, and to the publishers, particularly to Colin Greasby, for their kindness and patience at all times, as well as to Liz Greasby for her editorial work and for aiding in the compilation of the tables and prelims. The book endeavours to reflect the law as it was conceived to be in August 2000, when the text was prepared for press, although one or two further matters were included at proof stage.

PF Smith
Reading

Contents

Foreword iii
Preface v
Select bibliography xi
Table of cases xiii
Table of statutes xxxii

Chapter 1. Introduction

I – Overall review of aspects of the current law. II – Some statistics. III – The specific problem of dampness. IV – Implied covenant route. V – Interpreting express repairing obligations. VI – Reform of the law. VII – Extension of liability of residential landlords 1

Chapter 2. Waste

I – Introduction. II – Types of waste. III – Extent of liability for waste. IV – Remedies for waste. V – Reform of the law of waste 15

Chapter 3. Implied obligations of landlord and tenant as to repair and fitness

I – General principles. II – Implied obligations of landlord to repair. III – Warranties of fitness. IV – Implied obligations of tenant. V – Reform of the law 23

Chapter 4. Repairing obligations of landlords

I – General considerations. II – Extension of landlord liability. III – Dependent and independent covenants to repair. IV – Interpretation of landlords' express covenant. V – Requirement of notice of want of repair. VI – Landlord's right of entry to carry out repairs. 37

Chapter 5. Express repairing obligations of tenant

I – Introductory and general aspects. II – Construction of repairing obligations. III – Covenants against structural alterations to the premises. IV – Rent review implications 61

Chapter 6. Landlord's remedies for tenants' breaches of repairing obligations

I – Introduction. II – Impact of damages of Leasehold Property (Repairs) Act 1938. III – Section 18(1) of the Landlord and Tenant Act 1927. IV – Specific performance against the tenant 99

Chapter 7. Forfeiture for breach of covenant to repair

I – General principles. II – Status of lessee during forfeiture proceedings. III – Statutory notice requirements and relief against forfeiture 117

Chapter 8. Tenants' remedies for breach of covenant to repair

I – Repudiation of lease or tenancy. II – Damages for breach of landlord's covenant to repair. III – Deduction or set-off from rent. IV – Specific performance. V – Appointment of receiver and manager. VI – Control of service charges........ 144

Chapter 9. Insurance and reinstatement of premises

I – General principles. II – Further considerations 171

Chapter 10. Liability to repair imposed by statute

I – Introduction. II – Statute-implied repairing obligations of landlords of short residential tenants. III – Landlord's duty under Defective Premises Act 1972. IV – Incidence of liability for repairs in particular classes of residential leases. V – Effect of disrepair on rent and terms of new business tenancies. VI – Control of unfit housing by local authorities. VII – Control of statutory nuisances 179

Contents

Chapter 11. Party walls and dangerous structures

I – Background to statutory rules. II – Statutory regulation of rights of party wall and party fence wall owners. III – Dangerous buildings or structures 212

Chapter 12. Agricultural dilapidations

I – Introduction. II – Liability for repairs to buildings and fixed equipment under the rules of the 1986 Act. III – Damages claims by landlord under the 1986 Act. IV – Dilapidations and farm business tenancies 230

Chapter 13. Third party rights and liabilities

I – Introduction. II – Liability of landlords in nuisance. III – Liability of lessees in nuisance. IV – Liability of occupiers to lawful visitors and others. V – Access to neighbouring land ... 242

Chapter 14. Miscellaneous aspects

I – Covenants to repair freehold land. II – Compensation for business tenants' improvements. III – Rent review implications. IV – Effect of listed building controls. V – Options to break, renew or purchase. VI – Liability of tenant in respect of fixtures. 265

Index 283

Select bibliography

The following is a list of the principal works referred to in this book.

Auque, *Baux Commerciaux, Théorie et Pratique*, 1996, LGDJ, Paris
Bernstein and Reynolds, *Handbook of Rent Review* (looseleaf), Sweet & Maxwell, London
Bickford-Smith and Sydenham, *Party Walls, The New Law*, 1997, Jordans, Bristol
Cheshire and Burn's *Modern Law of Real Property*, 16th edn, 2000, Butterworths, London
Code Civil, 1997/98, Dalloz, Paris
Davey, *Landlord and Tenant Law*, 1999, Sweet & Maxwell, London
Evans and Smith, *The Law of Landlord and Tenant*, 5th edn, 1997, Butterworths, London
Fancourt, *Enforcement of Leasehold Covenants*, 1997, Sweet & Maxwell, London
Foa's General Law of Landlord and Tenant, 8th edn, 1957, Thames Bank, London
Gordon, *Scottish Land Law*, 1989, W Green & Son, Edinburgh
Hill & Redman's Law of Landlord and Tenant (looseleaf), Butterworths, London
Hollis and Gibson, *Surveying Buildings*, 3rd edn, 1991 (also 2nd edn, 1986), RICS Books, London
Lewison, *Drafting Business Leases*, 5th edn, 1996, FT Law and Tax, London
Luba and Knafler, *Repairs – Tenants' Rights*, 2nd edn, 1999, Legal Action Group, London
Luxton and Wilkie, *Commercial Leases*, 1998, CLT Professional Publishing, Birmingham
Marshall, Worthing and Heath, *Understanding Housing Defects*, 1998, Estates Gazette, London
McAllister, *Scottish Law of Leases*, 1989, Butterworths, London
Megarry and Wade, *The Law of Real Property*, 6th edn, 2000, Sweet & Maxwell, London
Melville and Gordon, *The Repair and Maintenance of Houses*, 2nd edn, 1997, Estates Gazette, London
Muir Watt and Moss, *Agricultural Holdings*, 14th edn, 1999, Sweet & Maxwell, London
Oxley and Gobert, *Dampness in Buildings*, 2nd edn, 1994, Butterworths Heinemann, Oxford

Paton and Cameron, *Landlord and Tenant*, 1967, W Green & Son, Edinburgh
Precedents for the Conveyancer (looseleaf), Sweet & Maxwell, London
Ross, *Commercial Leases*, 5th edn, 1998 (also 4th edn, 1994), Butterworths, London
Salmond and Heuston on Torts, 21st edn, 1996, Sweet & Maxwell, London
Scammell and Densham, *Agricultural Holdings*, 8th edn, 1996, Butterworths, London
Seeley, *Building Surveys, Reports and Dilapidations*, 1985, Houndmills, Basingstoke
Thornton, *Property Disrepair and Dilapidations*, 1992, Fourmat Publishing, London
Tromans, *Commercial Leases*, 2nd edn, 1996, Sweet & Maxwell, London
Woodfall, *Landlord and Tenant* (looseleaf), Sweet & Maxwell, London

Table of cases

A

Abbey National Building Society v *Maybeech Ltd* [1985] Ch 190......... 133
Adagio Properties Ltd v *Ansari* [1998] 2 EGLR 69, CA............. 127, 128
Adami v *Lincoln Grange Management Ltd* [1998] 1 EGLR 58, CA........ 26
Adams v *Rhymney Valley District Council* [2000] 39 EG 144, CA........ 31
AGB Research plc, Re [1994] EGCS 73............................. 118
Alcatel Australia Ltd v *Scarcella* (1998) 44 NSWLR 349............. 66, 83
Alexander v *Lambeth London Borough Council* [2000] 2 CL 386....... 55, 79
Alfred McAlpine Construction Ltd v *Panatown Ltd* The Times, August 15 2000, HL... 49
Al Hassani v *Merrigan* [1988] 1 EGLR 93, CA....................... 57
Alliance Economic Investment Co v *Berton* (1923) 92 LJKB 750.......... 91
Amsprop Trading Ltd v *Harris Distribution Ltd* [1997] 2 EGLR 78....... 102
Argy Trading Development Co v *Lapid Developments Ltd* [1977] 1 WLR 444... 171
Arnold (Paul) v *Greenwich London Borough Council* [1998] CLY 3618.... 146
Ashdown v *Samuel Williams & Sons Ltd* [1957] 1 QB 409, CA...... 254, 257
Associated British Ports v *CH Bailey plc* [1990] 1 EGLR 77, HL; affirming [1989] 1 EGLR 69, Ch.................... 130, 141, 142, 143
Associated Deliveries Ltd v *Harrison* [1984] 2 EGLR 76, CA 124
Austerberry v *Oldham Corporation* (1885) 29 ChD 750, CA............ 266
Austin v *Bonney* [1999] QdR 114 190

B

Bailey v *Armes* [1999] EGCS 21, CA............................. 249
Bairstow Eves (Securities) Ltd v *Ripley* [1992] 2 EGLR 47, CA.......... 278
Ballard (Kent) Ltd v *Oliver Ashworth (Holdings) Ltd* [1999] 2 EGLR 23... 121, 122
Balls Bros Ltd v *Sinclair* [1931] 2 Ch 325 96
Bank of Ireland Home Mortgages v *South Lodge Developments* [1996] 1 EGLR 91.. 135
Barker v *Herbert* [1911] 2 KB 633, CA............................ 248
Barnes v *Dowling* (1881) 44 LT 809............................... 20
Barrett v *Lounova (1982) Ltd* [1988] 2 EGLR 54, CA 6, 24
Barry v *Minturn* [1913] AC 584, HL............................. 222
Basingstoke and Deane Borough Council v *Host Group Ltd* [1987] 2 EGLR 147, CA 270

Bass Holdings Ltd v *Morton Music Ltd* [1987] 1 EGLR 214, CA 277, 278
Bath City Council v *Secretary of State for the Environment* [1983]
  JPL 737.. 276
Bavage v *Southwark London Borough Council* [1998] CLY 3623.......... 52
Beacon Carpets Ltd v *Kirby* [1984] 2 All ER 726..................... 177
Belcher v *M'Intosh* (1839) 2 M&R 186 76
Bennett v *Herring* (1857) 3 CB (NS) 370.............................. 89
Benzie v *Happy Eater Ltd* [1990] EGCS 76 213
Berkeley v *Poulett* [1977] 1 EGLR 86, CA 280
Berry v *Newport Borough Council* [2000] 29 EG 127, CA................ 48
Berrycroft Management Co Ltd v *Sinclair Gardens Investments
  (Kensington) Ltd* [1997] 1 EGLR 47, CA............................... 173
Betts (Frederick) Ltd v *Pickfords Ltd* [1906] 2 Ch 87 217
Bhojwani v *Kingsley Investment Trust Ltd* [1992] 2 EGLR 70........... 129
Bickmore v *Dimmer* [1903] 1 Ch 158................................... 90
Biggin Hill Airport Ltd v *Bromley London Borough Council* The Times,
  January 9 2001... 282
Billson v *Residential Apartments Ltd* [1992] 1 EGLR 43, HL; reversing
  [1991] 1 EGLR 70, CA 121, 129, 133
Billson v *Residential Apartments Ltd* [1993] EGCS 150............ 131, 132
Billson v *Tristrem* [2000] L&TR 220, CA......................... 160, 161
Bishop v *Consolidated London Properties Ltd* (1933) 102 LJKB 257..... 52, 76
Blewett v *Blewett* [1936] 2 All ER 188............................. 126
Boehm v *Goodall* [1911] 1 Ch 155 154
Boldack v *East Lindsay District Council* (1999) 31 HLR 41, CA 191
Boldmark Ltd v *Cohen* [1986] 1 EGLR 47 162
Boswell v *Crucible Steel Co* [1925] 1 KB 119, CA 45
Botham v *TSB Bank plc* (1997) 73 P&CR D1...................... 279, 280
Botross v *Hammersmith and Fulham London Borough Council* (1994)
  93 LGR 268... 211
Bounds v *Camden London Borough Council* [1999] CLY 3728............. 166
Boyd v *Wilton* [1957] 2 QB 277, CA 239
Boyle v *Tamlyn* (1827) 6 B&C 329................................... 268
Bradley v *Chorley Borough Council* [1985] 2 EGLR 49, CA......... 147, 183
Brent London Borough Council v *Carmel* (1996) 28 HLR 203, CA ... 146, 147
Brett v *Brett Essex Golf Club Ltd* [1986] 1 EGLR 154, CA 97
Brew Bros Ltd v *Snax (Ross) Ltd* [1970] 1 QB 612, CA 69, 79, 243, 247
Brikom Investments Ltd v *Carr* [1979] QB 467, CA..................... 37
Brikom Investments Ltd v *Seaford* [1981] 2 All ER 783, CA 182
British Anzani (Felixstowe) Ltd v *International Marine Management (UK)
  Ltd* [1979] 1 EGLR 65 150
British Telecom plc v *Sun Life Assurance Society plc* [1995] 2 EGLR
  44, CA... 52, 56, 63, 76, 182, 190
Brown v *Liverpool Corporation* [1969] 3 All ER 1345, CA 185
Bruton v *London & Quadrant Housing Trust* [1999] 2 EGLR 59, HL..... 181

Table of cases xv

Bryant v Foot (1867) LR 2 QB 161. 268
Brydon v Islington London Borough Council [1997] CLY 1754 146
Bunn v Harrison (1886) 3 TLR 146, CA 48
Burchell v Hornsby (1808) 1 Camp 360. 20
Burden v Hannaford [1956] 1 QB 142, CA 231
Burlington Property Co Ltd v Odeon Theatres Ltd [1938] 3 All ER
  469, CA 227
Buswell v Goodwin [1971] 1 WLR 92, CA. 35
Butuyuyu v Hammersmith and Fulham London Borough Council (1997)
  29 HLR 584 210

C

Calabar Properties Ltd v Stitcher [1983] 2 EGLR 46, CA 147
Calthorpe v McOscar [1924] 1 KB 716, CA 5, 64, 68, 72, 78, 84, 88
Cambridge Water Co v Eastern Counties Leather plc [1994] 2 AC
  264, HL. 244
Camden London Borough Council v Gunby [1999] 3 EGLR 13 208
Camden London Borough Council v London Underground Ltd [2000]
  Env LR 369 207
Camden Theatre v London Scottish Properties Ltd (Unreported,
  November 30 1984). 178
Campden Hill Towers v Gardner [1977] QB 823, CA. 188
Campden Hill Towers v Marshall (1965) 196 EG 989. 163
Cannock v Jones (1849) 3 Ex 233 50
Cantors Properties (Scotland) Ltd v Swears & Wells Ltd 1978 SC 310. 10
Capital & Counties Freehold Equity Trust Ltd v BL plc [1987] 2
  EGLR 49. 163
Carr v Hackney London Borough Council (1995) 93 LGR 606 .... 206, 209
Carstairs v Taylor (1871) LR 6 Ex 217. 28
Cavalier v Pope [1906] AC 428, HL. 2, 30, 191
Charsley v Jones (1889) 53 JP 280 32
Chartered Trust plc v Davies [1997] 2 EGLR 83, CA. 145
Chartered Society of Physiotheraphy v Simmons Church Smiles [1995]
  1 EGLR 155. .... 212, 219, 227
Chatfield v Elmstone Resthouse Ltd [1975] 2 NZLR 269 .... 57, 58
Chatham Empire Theatre (1955) Ltd v Ultrans Ltd [1961] 1 WLR 817 134
Chaytor v London, New York and Paris Association of Fashion and Price
  (1961) 30 DLR (2d) 527. 256
Cheetham v Hampson (1791) 4 TR 318 247
Chelsea Cloisters Ltd, Re (1980) 41 P&CR 98, CA. 163
Chelsea (Viscount) v Hutchinson [1994] 2 EGLR 48, CA 133
Chelsea (Viscount) v Muscatt [1990] 2 EGLR 48, CA 90
Chelsea Yacht & Boat Co Ltd v Pope [2000] 22 EG 147, CA 279, 280
Cheverall Estates Ltd v Harris [1998] 1 EGLR 27 103
Christian v Johanssen [1956] NZLR 664 256

Church of Our Lady of Hal v Camden London Borough Council [1980]
  2 EGLR 32, CA 204
Church Commissioners for England v Ve-Ri-Best Manufacturing Co [1957]
  1 QB 238... 141
CIN Properties Ltd v Barclays Bank plc [1986] 1 EGLR 59, CA 51
City of London Corporation v Fell [1994] 2 EGLR 131, HL.............. 63
City Offices plc v Bryanston Insurance Co Ltd [1993] 1 EGLR 126....... 270
Clarke v Findon Developments Ltd [1984] 1 EGLR 129................ 271
Clarke v Taff-Ely Borough Council (1980) 10 HLR 44................. 192
Clayton v Sale Urban District Council [1926] 1 KB 415 211
Clowes v Bentley Pty Ltd [1970] WAR 24.......................... 80
Cockburn v Smith [1924] 2 KB 119, CA 27
Collins v Flynn [1963] 2 All ER 1068 74
Connaught Restaurants Ltd v Indoor Leisure Ltd [1993] 2 EGLR
  108, CA 150
Cooke (ED & AD) Bourne (Farms) Ltd v Mellows [1982] 2 All ER
  208, CA 239
Co-operative Wholesale Society v National Westminster Bank plc [1995]
  1 EGLR 97, CA 271
Cornillie v Saha (1996) 72 P&CR 147, CA 122
Courage Ltd v Crehan [1999] 2 EGLR 145, CA................... 149, 150
Coventry City Council v Cartwright [1975] 2 All ER 99 209
Coventry City Council v Cole [1994] 1 EGLR 63, CA.................. 165
Coventry City Council v Doyle [1981] 1 WLR 1325 210
Craven (Builders) Ltd v Secretary of State for Health [2000]
  1 EGLR 128... 99, 104, 110
Crawford v Clarke [2000] EGCS 33, CA 131
Crawford v Newton (1886) 36 WR 54, CA 86
Credit Suisse v Beegas Nominees Ltd [1994] 1 EGLR 76 ... 7, 51, 67, 145, 146
Creery v Summersell [1949] Ch 751............................... 134
Cremin v Barjack Properties Ltd [1985] 1 EGLR 30, CA............... 131
Creska Ltd v Hammersmith and Fulham London Borough Council [1998]
  3 EGLR 35, CA 53, 69, 81
Crewe Services & Investment Corporation v Silk [1998] 2 EGLR
  1, CA 105, 230
Crofts v Haldane (1867) LR 2 QB 194 225
Crosby v Alhambra Co Ltd [1907] 1 Ch 295......................... 220
Crow v Wood [1971] 1 QB 77, CA................................ 268
Crown Estate Commissioners v Signet Group plc [1996)] 2 EGLR 200 130
Crown Estate Commissioners v Town Investments Ltd [1992] 1 EGLR
  61 .... 109, 113
Culworth Estates Ltd v Society of Licensed Victuallers [1991] 2 EGLR
  54, CA... 105, 107
Cunard v Antifyre Ltd [1933] 1 KB 551............................ 247
Cunliffe v Goodman [1950] 2 KB 237, CA 114

Table of cases xvii

Curtis v London Rent Assessment Committee [1997] 4 All ER 842, CA ... 196
Cusack-Smith v Gold [1958] 2 All ER 361 .... 141, 142

D

Daiches v Bluelake Investments Ltd [1985] 2 EGLR 67 154
Darcy (Lord) v Askwith (1617) Hob 234 17
Darlington Borough Council v Denmark Chemists Ltd [1993] 1 EGLR
  62, CA 131
Davies v Yadegar [1990] 1 EGLR 70, CA 95
Davstone Estates Ltd's Lease, Re [1969] 2 Ch 378 44
Dayani v Bromley London Borough Council [1999] 3 EGLR 144.......... 19
D'Eyncourt v Gregory (1866) LR 3 Eq 382 280
De Lassalle v Guildford [1901] 2 KB 215, CA 48
Dean v Walker (1996) 73 P&CR 366, CA 259
Debtors no 13A10 and 14A10 of 1994, In re [1995] 2 EGLR 33.......... 120
Deen v Andrews [1986] 1 EGLR 262............................... 281
Defries v Milne [1913] 1 Ch 98, CA 15, 16
Demetriou v Poolaction Ltd [1991] 1 EGLR 100, CA.................. 6, 25
Department of Transport v Egoroff [1986] 1 EGLR 89, CA 183
Devonshire Reid Properties Ltd v Trenaman [1997] 1 EGLR 45........... 53
Dialworth Ltd v TG Organisation (Europe) Ltd (1996) 75 P&CR
  147, CA 28
Dickinson v Enfield London Borough Council [1996] 2 EGLR 88 271
Dickinson v St Aubyn [1944] 1 All ER 370 86
Digby v Atkinson (1815) 4 Camp 275 174
Dinefwr Borough Council v Jones [1987] 2 EGLR 58, CA................ 57
Doe d Grubb v Earl of Burlington (1833) 5 B&Ald 507................. 19
Doe d Worcester Trustees v Rowlands (1841) 9 C&P 734............. 88, 105
Doe d Pittman v Sutton (1841) 9 C&P 706 67
Doe v Withers (1831) 2 B&Ald 896................................ 23
Doherty v Allman (1878) 3 App Cas 709, HL............... 16, 18, 19, 20
Donegal Tweed Co v Stephenson (1929) 98 LJKB 657 269
Dover District Council v Farrar (1982) 2 HLR 32..................... 210
Drummond v S&U Stores Ltd [1981] 1 EGLR 42 .... 104, 108
Drury v Army and Navy Auxiliary Co-Operative Society Ltd [1896]
  2 QB 721... 217
Dun & Bradstreet Software Services (England) Ltd v Provident Mutual
  Life Assurance Association [1998] 2 EGLR 175, CA 278
Dunster v Hollis [1918] 2 KB 795 27
Durley House Ltd v Cadogan [2000] 1 WLR 246 273

E

Ebbetts v Conquest [1895] 2 Ch 377, CA........................... 112
Ecclesiastical Commissioners v Merrall (1869) LR 4 Ex 162 63
Edler v Auerbach [1950] 1 KB 359................................. 23

Edmonton Corporation v WM Knowles & Son Ltd (1961) 60 LGR 124 27
Egerton v Harding [1974] 3 All ER 689, CA 268
Electricity Supply Nominees Ltd v IAF Group Ltd [1993] 2 EGLR 95 151
Electricity Supply Nominees Ltd v National Magazine Co Ltd [1999]
  1 EGLR 130 148
Elite Investments Ltd v Bainbridge (TI) Silencers Ltd [1986] 2
  EGLR 43 .... 67, 79
Elite Investments Ltd v Bainbridge (TI) Silencers Ltd (No 2) [1987]
  2 EGLR 50 104
Elitestone Ltd v Morris [1997] 2 EGLR 115, HL 279, 280
Ellenborough Park, Re [1956] Ch 131, CA 267
Eller v Grovecrest Investments Ltd [1994] 2 EGLR 45, CA 150
Elmcroft Developments Ltd v Tankersley-Sawyer [1984] 1 EGLR
  47, CA 5, 55, 81
Empson v Forde [1990] 1 EGLR 131, CA 195
Escalus Properties Ltd v Robinson [1995] 2 EGLR 23, CA 135, 136
Espir v Basil Street Hotel Ltd [1936] 3 All ER 91, CA 112
Estates Projects Ltd v Greenwich London Borough Council [1979]
  2 EGLR 85 98
Euston Centre Properties Ltd v Wilson (H&J) Ltd [1982]1 EGLR 57 97
Evans v Jones [1955] 2 QB 58, CA 234, 236, 237, 240
Evans v Clayhope Properties Ltd [1988] 1 EGLR 33, CA 154
Exchange Travel Agency Ltd v Triton Property Trust plc [1991]
  2 EGLR 50 120
Expert Clothing Service & Sales Ltd v Hillgate House Ltd [1985] 2
  EGLR 85, CA 89, 122
Eyre v McCracken (2000) 80 P&CR 220, CA 68, 80
Eyre v Rea [1947] KB 567 110

F

Fairclough (TM) & Sons Ltd v Berliner [1931] 1 Ch 60 113
Family Management v Gray [1980] 1 EGLR 46 .... 108, 201
Farimani v Gates [1984] 2 EGLR 66, CA 171
Fawke v Chelsea (Viscount) [1979] 1 EGLR 89, CA 201
Ferguson v Welsh [1987] 1 WLR 1553, HL 253
Field v Curnick [1926] 2 KB 374 89
Fillingham v Wood [1891] 1 Ch 51 216
Filross Securities Ltd v Midgeley [1998] 3 EGLR 43 150
Fincar SRL v 109/113 Mount Street Management Co Ltd [1999] L&TR
  161, CA 67
Finch v Underwood (1876) 2 ChD 310, CA 278
Finchbourne Ltd v Rodrigues [1976] 3 All ER 581, CA 44, 161
Firstcross Ltd v Teasdale [1983] 1 EGLR 87 196
Fletcher v Nokes [1897] 1 Ch 271 127
Fox v Jolly [1916] 1 AC 1, HL 127, 128

Table of cases   xix

Francis v Cowlcliffe (1976) 33 P&CR 368 152
Fuller v Judy Properties Ltd [1992] 1 EGLR 75, CA 127

G
Gardner v Blaxill [1960] 1 WLR 752 279
Gething v Evans [1997] CLY 1753 146
Gibbs Mew plc v Gemmell [1999] 1 EGLR 43, CA 150
Gordon v Selico Co Ltd [1986] 1 EGLR 71, CA; [1985] 2 EGLR 79 151
Goreley, Ex parte (1914) LJ Bkcy 1 178
Gott v Gandy (1853) 2 El&B 845 26
Graham v Markets Hotel Pty Ltd (1943) 67 CLR 567 61
Granada Theatres Ltd v Freehold Investments (Leytonstone) Ltd [1959]
  Ch 592, CA; [1958] 1 WLR 845 .... 46, 60
Grayless v Watkinson [1990] 1 EGLR 6, CA 234
GREA Real Property Investments v Williams [1979] 1 EGLR 121 98
Greater London Council v Tower Hamlets London Borough Council (1983)
  15 HLR 57 209
Green v Superior Court of City and County of San Francisco (1974) 111
  Cal Rptr 704 70
Greenwich London Borough Council v Discreet Selling Estates Ltd
  [1990] 2 EGLR 65, CA 122, 127
Griffin v Pillett [1926] 1 KB 17 58
Guillemard v Silverthorne (1908) 99 LT 584 128
Gutteridge v Munyard [1834] 1 M&R 334 80
Gyle-Thompson v Wall Street (Properties) Ltd [1974] 1 All ER 295 ... 218, 227

H
Habinteg Housing Association v James (1995) 27 HLR 299 .... 1, 207
Hafton Properties Ltd v Camp [1994] 1 EGLR 67 25
Hagee (London) Ltd v Co-operative Insurance Society [1992] 1 EGLR 57 ... 94
Haines v Florensa [1990] 1 EGLR 73, CA 95
Hall v Howard [1988] 2 EGLR 75, CA 57
Halliard Property Co Ltd v Clarke (Nicholas) Investments Ltd [1984] 1
  EGLR 45 79
Hallinan v Jones [1984] 2 EGLR 20 240
Hallisey v Petmoor Developments Ltd [2000] EGCS 124 .... 34, 45
Halsall v Brizell [1957] Ch 169 267
Hamilton v Martell Securities [1984] Ch 266 102
Hamish Cathie Travel England Ltd v Insight International Tours Ltd
  [1986] 1 EGLR 244 97
Hammersmith and Fulham London Borough Council v Creska (No 2)
  [2000] L&TR 288 59
Hammersmith and Fulham London Borough Council v Monk [1992] 1
  EGLR 65, HL 181
Hammond v Allen [1994] 1 All ER 307; [1993] 1 EGLR 1 .... 232, 233

Hanson v Newman [1934] Ch 298, CA 104
Hargroves, Aronson & Co v Hartopp [1905] 1 KB 472.................. 28
*Harmsworth Pension Funds Tustees Ltd v Charringtons Industrial
   Holdings Ltd* [1985] 1 EGLR 97................................ 271
Harnett v Maitland (1847) 16 M&W 257 20
Harrison v Thanet District Council [1998] CLY 3918 254
Hart v Emelkirk Ltd [1983] 3 All ER 15........................... 154
Hart v Windsor (1844) 12 M&W 68.......................... 5, 23, 26
Haskell v Marlow [1928] 2 KB 45 84
Havenridge Ltd v Boston Dyers Ltd [1994] 2 EGLR 73 .... 172, 173
Haviland v Long [1952] 2 QB 80, CA 107
Heap v Ind Coope & Allsop Ltd [1940] 2 KB 476, CA................. 242
Herne v Bembow (1813) 4 Taunt 764.............................. 18
Hewitt v Rowlands [1924] WN 135, CA 148
Hibernian Property Co Ltd v Liverpool Corporation [1973] 2 All ER 1117.. 114
Highway Properties Ltd v Kelly, Douglas & Co Ltd (1971) 17 DLR
   (3rd) 710.. 144
Hill v Barclay (1810) 16 Ves 402 114
Hill v Griffin [1987] 1 EGLR 81, CA.............................. 134
Hilton v Ankesson (1872) 27 LT 519 267
Historic Houses Hotels Ltd v Cadogan Estates [1995] 1 EGLR 117, CA;
   [1993] 2 EGLR 151, Ch 97
Hogarth Health Club Ltd v Westbourne Investments Ltd [1990] 1 EGLR
   89, CA 269
Holding & Barnes plc v Hill House Hammond Ltd [2000] L&TR 428 ... 25, 68
Holding & Management Ltd v Property Holding & Investment Trust plc
   [1990] 1 EGLR 65, CA....................................... 44, 68
Holiday Fellowship v Viscount Hereford [1959] 1 WLR 211.............. 45
Holland v Hodgson (1872) LR 7 CP 328 280
Holme v Crosby Corporation (Unreported, 1941) 229
Honywood v Honywood (1874) LR 18 Eq 306 17
Hopley v Tarvin Parish Council (1910) 74 JP 209...................... 129
Hopwood v Cannock Chase District Council [1975] 1 WLR 373, CA...... 185
Howard v Midrome Ltd [1991] 1 EGLR 58 156
Hudson v Williams (1878) 39 LT 632.............................. 88
Hungerford (Dame Margaret) Charity Trustees v Beazeley [1993]
   2 EGLR 143 .... 82, 180
Hunt v Harris (1865) 19 CB NS 13 216
Hussein v Mehlman [1992] 2 EGLR 87.............. 47, 145, 182, 190, 191
Hyman v Rose [1912] AC 623, HL.............................. 19, 130
Hynes v Twinsectra Ltd [1995] 2 EGLR 69, CA................... 124, 129

I

Iceland Frozen Foods plc v Starlight Investments Ltd [1992] 1
   EGLR 126, CA.. 273

Table of cases xxi

Independent Tank Cleaning v *Zabokrzeki* (1997) 8 RPR (3d) 177 174
Inderwick v *Leach* (1885) 1 TLR 484 113
Iperion Investments Corporation v *Broadwalk House Residents Ltd*
　[1992] 2 EGLR 235 .... 91, 123, 132
Ipswich Town Football Club Co Ltd v *Ipswich Borough Council* [1988]
　2 EGLR 146 97
Irvine v *Moran* [1991] 1 EGLR 261 .... 45, 183, 184, 185
Irving v *London County Council* (1965) 109 SJ 157 251
Issa v *Hackney London Borough Council* [1997] 1 All ER
　999 .... 29, 191, 204, 207
Ivory Gate Ltd v *Capital City Leisure Ltd* [1993] EGCS 76 273
Ivory Gate Ltd v *Spetale* [1998] 2 EGLR 43, CA 123
IVS Enterprises Ltd v *Chelsea Cloisters Management Ltd* [1994] EGCS
　14, CA 161, 165

J
J (A minor) v *Staffordshire County Council* [1997] CLY 3783 251
Jacobs v *London County Council* [1950] AC 361, HL 246
James v *Hutton and J Cook & Sons Ltd* [1950] 1 KB 9 .... 103, 110
Jaquin v *Holland* [1960] 1 All ER 402, CA 108, 184
Javins v *First National Realty Corpn* (1970) 428 F 2d 1071 .... 12, 70
Jeffs v *West London Property Corporation* [1954] CLY 1807 111
Jervis v *Harris* [1996] 1 EGLR 78, CA 59, 100, 101, 102
Jeune v *Queens Cross Properties Ltd* [1974] Ch 97 152
Jolley v *Sutton London Borough Council* [2000] 1 WLR 1082 .... 250, 252
Jollybird Ltd v *Fairzone Ltd* [1990] 2 EGLR 55, CA 161
Jones v *Herxheimer* [1950] 2 KB 106, CA 106
Jones v *Price* [1965] 2 QB 618, CA 268
Jones v *Pritchard* [1908] 1 Ch 630 213
Joseph v *London County Council* (1914) 111 LT 276 90
Josephine Trust v *Champagne* [1963] 2 QB 160, CA 91
Joyce v *Liverpool City Council* [1995] 3 All ER 110, CA 152
Joyner v *Weeks* [1891] 2 QB 31, CA 103

K
Kammins Ballrooms Ltd v *Zenith Investments (Torquay) Ltd* [1971]
　AC 850, HL 140
Kanda v *Church Commissioners for England* [1958] 1 QB 332 141
Kausar v *Eagle Star Insurance Co Ltd* The Times, July 15 1996, CA 174
Keats v *Graham* [1959] 3 All ER 919, CA 114
Kelly v *Woolworth & Co* [1922] 2 IR 5 249
Kemra (Management) Ltd v *Lewis* [1999] CLY 3729 149
Kenny v *Kingston-upon-Thames Royal London Borough* [1985] 1 EGLR
　26, CA 205
Kent v *Coniff* [1953] 1 QB 361, CA 238

Kiddle v City Business Properties Ltd [1942] 1 KB 269 28
King, Re [1963] Ch 459, CA; reversing [1962] 1 WLR 632 .... 109, 176
King v South Northamptonshire District Council [1992] 1 EGLR 53,
  CA 185
Kinlyside v Thornton (1776) 2 Blw 1111 16
Kirklinton v Wood [1917] 1 KB 332 85
Kleinwort Benson Ltd v Lincoln City Council [1998] 3 WLR
  1095 .... 43, 158, 159
Knight v Purcell (1879) 11 ChD 412 .... 213, 217

L

Ladbroke Hotels Ltd v Sandhu [1995] 2 EGLR 92 .... 65, 82
Lambert v FW Woolworth & Co Ltd (No 2) [1938] Ch 883, CA 95
Lambert v Keymood Ltd [1997] 2 EGLR 70 174
Lambeth London Borough Council v Rogers [2000] 1 EGLR 28, CA .. 124, 181
Lambeth London Borough Council v Stubbs (1980) 78 LGR 650 210
Land Securities plc v Westminster City Council (No 2) [1995] 1 EGLR
  245, CA 87
Landeau v Marchbank [1949] 2 All ER 172 110
Lansdowne Rodway Estates Ltd v Potown Ltd [1984] 2 EGLR 80 105
Larksworth Investments v Temple House Ltd (No 2) [1999] BLR
  297, CA 145, 146
Laura Investment Co Ltd v Havering London Borough Council [1992]
  1 EGLR 155 .... 97, 271
Leadbetter v Marylebone Corporation [1905] 1 KB 771, CA............. 219
Leanse v Egerton [1943] KB 323 246
Lee-Parker v Izzet [1971] 3 All ER 1099 149
Lehmann v Herman [1993] 1 EGLR 172 216
Lewis (John) Properties plc v Viscount Chelsea [1993] 2 EGLR
  77 .... 89, 113, 123
Lex Service plc v Johns [1990] 1 EGLR 92, CA 125
Lilley & Skinner Ltd v Crump (1929) 73 SJ 366 96
Lister v Lane and Nesham [1893] 2 QB 212 .... 70, 74, 79
Liverpool City Council v Irwin [1977] AC 239, HL 24, 186
Liverpool Properties Ltd v Oldbridge Investments Ltd [1985] 2
  EGLR 111 .... 16, 124
Lloyds Bank Ltd v Lake [1961] 1 WLR 884 112
Loader v Kemp (1826) 2 C&P 375 174
Lomax Leisure Ltd, In re [1999] 2 EGLR 37 120
London & Leeds Estates Ltd v Paribas [1993] 2 EGLR 149 272
London & Manchester Assurance Co Ltd v O & H Construction Ltd
  [1989] 2 EGLR 185 218
London Borough of Newham v Patel (1978) 13 HLR 77, CA 179
London County Council v Hutter [1925] Ch 626 90
London County Council v Jones [1912] 2 KB 504 229

Table of cases xxiii

London County Freehold and Leasehold Properties Ltd v Wallis-Whiddett
[1950] WN 180.. 109
London, Gloucestershire and North Hants Dairy Co v Morley and
Lanceley [1911] 2 KB 257 217
London Underground Ltd v Shell International Petroleum Co Ltd [1998]
EGCS 97... 62
Lonsdale & Thompson Ltd v Black Arrow Group plc [1993] 1 EGLR 87 ... 173
Loria v Hammer [1989] 2 EGLR 249 .... 45, 149
Lotteryking Ltd v AMEC Properties Ltd [1995] 2 EGLR 13.............. 48
Louis v Sadiq [1997] 1 EGLR 136, CA............................. 221
Lovelock v Margo [1963] 2 QB 786, CA............................ 95
Lowe v Quayle Munro Ltd 1997 SLT 1168.......................... 10
Lurcott v Wakely and Wheeler [1911] 1 KB 905, CA 68, 78

M

McCall's Entertainments (Ayr) Ltd v South Ayrshire Council (No 2)
1998 SLT 1421 10
McAuley v Bristol City Council [1991] 2 EGLR 64, CA 26, 193
McCarrick v Liverpool Corporation [1947] AC 219, HL.............. 57, 190
McCoy & Co v Clark (1982) 13 HLR 87, CA........................ 147
McCulloch v Elsholz (Unreported, January 16 1990).................. 95
McDougall v Easington District Council [1989] 1 EGLR 93,
CA... 53, 60, 72, 187, 189
McGinley v British Railways Board [1983] 1 WLR 1427, HL 250
M'Glone v British Railways Board 1966 SC (HL) 1................... 250
McGreal v Wake [1984] 1 EGLR 42, CA.............. 57, 58, 147, 179, 191
McMullen & Sons Ltd v Cerrone [1994] 1 EGLR 99 120
McNerny v Lambeth London Borough Council [1989] 1 EGLR
81, CA... 1, 30, 180, 192
Mancetter Developments Ltd v Garmanson Ltd [1986] 1 EGLR
240, CA... 15, 16, 17, 20
Manchester v Dixie Cup Co (1952) 1 DLR 19 52
Manchester Bonded Warehouse Co v Carr (1880) 5 CPD 507 18
Mannai Insurance Co Ltd v Eagle Star Life Assurance Co Ltd [1997]
AC 749... 68
Manor House Drive Ltd v Shahbazian (1965) 195 EG 283.............. 162
Marchant v Capital & Counties plc [1983] 2 EGLR 156, CA............ 229
Marenco v Jacramel Co Ltd (1964) 191 EG 433 161
Marfield Properties Ltd v Secretary of State for the Environment 1996
SLT 1244.. 2
Marker v Kenrick (1853) 13 CB 188................................ 15
Marlton v Turner [1997] CLY No 4233 266
Marsden v Edward Heyes Ltd [1927] 2 KB 1, CA 19, 33
Martin v Maryland Estates Ltd [1999] 2 EGLR 53, CA............ 166, 167
Maryland Estates Ltd v Bar Joseph [1998] 2 EGLR 47, CA 62, 123

Masterton Licensing Trust v *Finco* [1957] NZLR 1137, CA 54
Mather v *Barclays Bank plc* [1987] 2 EGLR 254 107
Matthey v *Curling* [1922] 2 AC 180, HL........................... 175
Meadows v *Clerical, Medical and General Life Assurance Society* [1981]
  Ch 70 124
Melville v *Grapelodge Developments Ltd* [1980] 1 EGLR 42 .... 149, 150
Metropolitan Fim Studio's Ltd's Application, Re [1962] 1 WLR 1315 143
Meux v *Cobley* [1892] 2 Ch 253................................... 18
Mickel v *M'Coard* 1913 SC 896 34
Middlegate Properties Ltd v *Gidlow-Jackson* (1977) 34 P&CR 4, CA...... 136
Middlegate Properties Ltd v *Messimeris* [1973] 1 WLR 168, CA 140
Miller v *Burt* (1918) 63 SJ 117 83
Minja Properties Ltd v *Cussins Property Group plc* [1998] 2 EGLR
  52, CA.. 54, 55, 65, 69, 71
Mint v *Good* [1951] 1 KB 517, CA........................ 243, 244, 245
Mira v *Alymer Square Investments Ltd* [1990] 1 EGLR 45, CA 148
Monk v *Noyes* (1824) 1 C&P 265................................... 86
Morcom v *Campbell-Johnson* [1955] 1 QB 106, CA 68, 72, 73
Morris v *Liverpool City Council* [1988] 1 EGLR 47, CA................ 58
Moss' Empires Ltd v *Olympia (Liverpool) Ltd* [1939] AC 544, HL........ 110
Mount Cook Land Ltd v *Hartley* [2000] EGCS 26 130
Mullaney v *Maybourne Grange (Croydon) Management Co Ltd* [1986]
  1 EGLR 70 .... 54, 65
Mumford Hotels Ltd v *Wheler* [1964] Ch 117 178
Musgrove v *Pandelis* [1919] 2 KB 43.............................. 178

N

National Carriers Ltd v *Panalpina (Northern) Ltd* [1981] AC 675,
  HL 90, 175
National Real Estate & Finance Co Ltd v *Hassan* [1939] 2 KB 61 .... 100, 139
N&D (London) Ltd v *Gadsdon* [1992] 1 EGLR 112 197
NCB v *Thorne* [1976] 1 WLR 543 209
Network Housing Association v *Westminster City Council* (1994) 93
  LGR 280.. 209
New England Properties Ltd v *Portsmouth New Shops* [1993] 1 EGLR
  84 .... 7, 55
New Zealand Government Property Corpn v *HM&S Ltd* [1982] QB 1145. . 272
Nind v *Nineteenth Century Building Society* [1894] 2 QB 226, CA....... 136
Noble v *Harrison* [1926] 2 KB 332............................... 248
Northways Flats Management Co (Camden) Ltd v *Wimpey Pension
  Trustees Ltd* [1992] 2 EGLR 42, CA................................ 51
Norwich Union Life Assurance Society v *British Railways Board* [1987]
  2 EGLR 137 .... 7, 66
Nurdin & Peacock plc v *DB Ramsden & Co Ltd (No 2)* [1999] 1
  EGLR 15.. 158

Table of cases   xxv

Nynehead Developments Ltd v RH Fibreboard Containers Ltd [1999] 1 EGLR 7 145

O

O'Brien v Robinson [1973] AC 912, HL 58, 182, 190
O'Leary v Meiltides and Eastern Trust Co (1960) 20 DLR 2d 258 244
O'May v City of London Real Property Co Ltd [1983] 2 AC 726, HL 201
Oakley v Birmingham City Council The Times, November 29 2000 2
Ocean Accident and Guarantee Corp v Next plc [1996] 2 EGLR 84 272
Official Custodian for Charities v Mackey [1985] Ch 168 134
Official Custodian for Charities v Mackey (No 2) [1985] 1 EGLR 46, CA 135
Old Grovebury Manor Farm v W Seymour Plant Sales & Hire Ltd (No 2) [1979] 3 All ER 504 126
Olympia & York Canary Wharf Ltd, Re [1993] BCLC 154 120

P

P&O Property Holdings Ltd v International Computers Ltd [1999] 2 EGLR 17 .... 62, 160
Pakwood Transport v 15 Beauchamp Place (1977) 36 P&CR 112, CA 126
Pan Australian Credits (SA) Pty Ltd v Kolim Pty Ltd (1981) 27 SASR 353 280
Pannell v City of London Brewery Co [1900] 1 Ch 496 128
Park v J Jobson & Son [1945] 1 All ER 222, CA 268
Parker v O'Connor [1974] 3 All ER 257, CA 183
Parker v Camden London Borough Council [1985] 2 All ER 141, CA 154
Passley v London Borough of Wandsworth (1996) 30 HLR 165, CA 52, 56, 76, 146, 190
Pearlman v Keepers and Governors of Harrow School [1979] 1 All ER 365, CA 73
Pellicano v MEPC plc [1994] 1 EGLR 104 135
Pemberton v Bright [1960] 1 WLR 436 247
Pemberton v Southwark London Borough Council [2000] 2 EGLR 33, CA 181
Pembery v Lamdin [1940] 2 All ER 434, CA 5, 54, 73, 80
Peninsular Maritime Ltd v Padseal [1981] 2 EGLR 43, CA 124
Penn v Gatenex Co Ltd [1958] 2 QB 210, CA 186
Pennial v Harborne (1848) 1 QB 368 171
Penton v Barnett [1898] 1 QB 276, CA 122
Perry v Chotzner (1893) 9 TLR 488 85
Pertemps Group Ltd v Crosher & James [1999] CLY 3676 136
Phillips v Mobil Oil Co [1989] 2 EGLR 246 282
Phillips v Price [1959] Ch 181 142
Phipps v Pears [1965] 1 QB 76, CA 224
Phipps v Rochester Corporation [1955] 1 QB 450 252

Pike v Sefton Metropolitan Borough Council [2000] EHLR Dig 272 210
Plough Investments Ltd v Manchester City Council [1989] 1
  EGLR 244 .... 39, 45, 53, 72, 74, 83, 102
Pole Properties Ltd v Feinberg (1981) 43 P&CR 121, CA 162
Pollway Nominees Ltd v Croydon London Borough Council [1987] AC
  79, HL 204
Polychronakis v Richards & Jerron Ltd [1998] Env LR 347 210
Ponsford v HMS Aerosols Ltd [1979] AC 63, HL 96
Pontsarn Investments Ltd v Kansallis-Osake-Pankki [1992] 1 EGLR
  148 272
Portman v Latta [1942] WN 97 109
Post Office v Aquarius Properties Ltd [1987] 1 EGLR 40,
  CA. 4, 55, 71, 74, 80
Powys v Blagrave (1854) 4 De GM&G 488 20
Pretty v Bickmore (1873) LR 8 CP 401 246
Pritchard v Peto [1917] 2 KB 173 248
Progessive Mailing House Pty Ltd v Tabali Pty Ltd (1985) 157 CLR 17 ... 144
Proudfoot v Hart (1890) 25 QBD 42, CA 76, 86, 87, 183
Prudential Assurance Co v Waterloo Real Estate Inc The Times, May
  13 1998 217

Q
Quick v Taff-Ely Borough Council [1985] 2 EGLR 50,
  CA 1, 29, 64, 71, 74, 78, 179, 187

R
R v Bristol City Council, ex parte Everett [1999] 3 PLR 14, CA 209
R v Falmouth and Truro Port Health Authority, ex parte South West Water
  Services Ltd [2000] 3 All ER 307, CA 208
R v Forest of Dean District Council The Times, November 9 1989 204
R v Leominster District Council, ex parte Antique Country Buildings
  Ltd [1988] JPL 554 276
R v London Borough of Southwark, ex parte Cordwell (1993) 26 HLR
  107 202
R v London Leasehold Valuation Tribunal, ex parte Daejan Properties Ltd
  [2000] 49 EG 121 166
R v McCarthy & Stone (Developments) Ltd [1998] CLY 4198 274
R v Recorder of Bolton [1940] 1 KB 290, CA 229
R v Sandhu (Major) [1997] Crim LR 288 274
R v Stroud District Council [1982] JPL 246 228
Rae v Mars (UK) Ltd [1990] 1 EGLR 161 252
Rainbow Estates Ltd v Tokenhold Ltd [1998] 2 EGLR 34 .... 100, 115
Rakhit v Carty [1990] 2 EGLR 95, CA 195
Ratcliff v McConnell [1999] 1 WLR 670 .... 255, 256
Ravenseft Properties Ltd v Davstone (Holdings) Ltd [1980] QB 12 .. 73, 74, 78

Table of cases xxvii

Rawlings v Morgan (1865) 18 CB (NS) 776 113
Redmond v Dainton [1920] 2 KB 256 .... 62, 175
Reed Personnel Services plc v American Express Ltd [1997] 1 EGLR
  229 277
Regional Properties Ltd v City of London Real Property Co Ltd [1981]
  1 EGLR 33 .... 59, 115, 152
Regis Property Co v Dudley [1959] AC 370, HL 84
Reid v Smith (1905) 3 CLR 656 280
Reste Realty Corporation v Cooper (1969) 33 ALR 3d 1341 8
Reston Ltd v Hudson [1990] 2 EGLR 51 .... 46, 146
Rexhaven Ltd v Nurse and Alliance and Leicester Building Society (1995)
  28 HLR 241 117
Rhone v Stephens (Executrix) [1994] 2 EGLR 181, HL 13, 213, 265, 266
Rich Investments Ltd v Camgate Litho Ltd [1988] EGCS 132 79
Riley Gowler Ltd v National Heart Hospital Board of Governors [1969]
  3 All ER 1401, CA 226
Rimmer v Liverpool City Council [1984] 1 EGLR 23, CA 31
Robbins v Jones (1863) 15 CB (NS) 221 30
Robbins v Secretary of State for the Environment [1989] 1 WLR 201 277
Rogan v Woodfield Building Services Ltd [1995] 1 EGLR 72 168
Roles v Nathan [1963] 1 WLR 1117 .... 252, 253
Rookes v Barnard [1964] AC 1129 20
Ropemaker Properties Ltd v Noonhaven Ltd [1989] 2 EGLR 50 132
Roper v Prudential Assurance Co Ltd [1992] 1 EGLR 5 234
Rowlands (Mark) Ltd v Berni Inns Ltd [1985] 2 EGLR 92, CA 174
Rugby School (Governors) v Tannahill [1935] 1 KB 87, CA 127
Rush v Lucas [1910] 1 Ch 437 17

S

Sack v Jones [1925] Ch 235 214
St Anne's Well Brewery Co v Roberts (1928) 44 TLR 703, CA 247
Salford City Council v McNally [1975] 2 All ER 860, HL 209
Salisbury v Gilmore [1942] 2 KB 38, CA 114
Sampson v Hodson-Pressinger [1982] 1 EGLR 50, CA 243
Saner v Bilton (1878) 7 ChD 815 60
Sarson v Roberts [1895] 2 QB 395, CA 32
Sarum Properties Ltd's Application, Re [1999] 2 EGLR 131 166
Scales v Lawrence (1860) 2 F&F 289 85
Scottish Mutual Assurance plc v Jardine Public Relations Ltd [1999]
  EGCS 43 161
Seaforth Land Sales (No 2), Re [1977] QdR 317 258
Secretary of State for the Environment v Euston Centre Investments
  (No 2) [1995] CLY 3062 271
SEDAC Investments Ltd v Tanner [1982] 3 All ER 646 100
Sedleigh-Denfield v O'Callaghan [1940] AC 880, HL 244, 245

Segal Securities Ltd v *Thoseby* [1963] 1 QB 887..................... 122
Selby v *Whitbread & Co* [1917] 1 KB 736.......................... 217
Sella House Ltd v *Mears* [1989] 1 EGLR 65, CA..................... 162
Service Oil Co Inc v *White* (1973) 242 P2d 652....................... 8
Shane v *Runwell* [1967] EGCD 88................................ 110
Sheldon v *West Bromwich Corporation* (1973) 25 P&CR 360............ 186
Shenkman Corpn v *OAC Holdings Ltd* (1997) 8 RPR 3d 118............ 55
Shepherd v *Lomas* [1963] 1 WLR 962, CA.......................... 235
Short v *Kirkpatrick* [1982] 2 NZLR 358............................ 280
Shortlands Investments Ltd v *Cargill plc* [1995] 1 EGLR
51.. 87, 106, 107, 281
Shrewsbury's (Countess of) Case (1600) 5 Co Rep 13b................. 20
Sidnell v *Wilson* [1966] 2 QB 67, CA.............................. 140
Silvester v *Ostrowska* [1959] 1 WLR 1060.......................... 128
Simmons v *Norton* (1831) 7 Bing 640............................... 17
Simpson v *Scottish Union Insurance Co* (1863) 1 H&M 618............ 178
Slater v *Worthington's Cash Stores (1930) Ltd* [1941] 1 KB 488......... 245
Sleafer v *Lambeth Borough Council* [1960] 1 QB 43, CA................ 26
Smedley v *Chumley & Hawke Ltd* [1982] 1 EGLR 47, CA............... 47
Smiley v *Townshend* [1950] 2 KB 311, CA...................... 105, 111
Smith v *Bradford Metropolitan Council* (1982) 44 P&CR 177....... 191, 193
Smith v *Marrable* (1843) 11 M&W 5............................... 31
Smith v *Metropolitan City Properties Ltd* [1986] 1 EGLR 52............ 124
Smith v *Mills* (1899) 16 TLR 59 89
Smith v *Thackerah* (1866) LR 1 CP 564............................ 214
Southern Depot Co Ltd v *British Railways Board* [1990] 2 EGLR 39
.... 130, 132
Southwark London Borough Council v *Mills* [1999] 3 EGLR 35, HL..... 5, 6,
.... 23, 24, 25, 29, 38, 243, 244
Spyer v *Phillipson* [1931] 2 Ch 183................................ 280
Standard Bank of British South America (Africa) v *Stokes* (1878)
9 ChD 68.. 213
Stanley v *Ealing London Borough Council* (2000) 32 HLR 745.......... 207
Staples v *West Dorset District Council* (1995) 93 LGR 536........ 253, 254
Starrokate Ltd v *Burry* [1983] 1 EGLR 56.................... 56, 128, 138
Staves v *Leeds City Council* [1992] 2 EGLR 37, CA................... 185
Stent v *Monmouth District Council* [1987] 1 EGLR 59,
CA.. 69, 81, 179, 188
Stocker v *Planet Building Society* (1879) 27 WR 877, CA............... 59
Straudley Investments Ltd v *Barpress Ltd* [1987] 1 EGLR 69............ 62
Sturolson & Co v *Mauroux* [1988] 1 EGLR 66, CA................... 146
Summers v *Salford Corporation* [1943] AC 283, HL.................. 203
Sutton v *Temple* (1843) 12 M&W 52............................... 28
Sutton (Hastoe) Housing Association v *Williams* [1998] 1 EGLR 56,
CA 44, 65

Table of cases xxix

Swain v Natui Ram Puri [1996] CLY 5697, CA 255
Swallow Securities Ltd v Brand [1981] 2 EGLR 48 101
Switzer v Law [1998] CLY 3624. 147

T

Target Home Loans Ltd v Iza Ltd [2000] 1 EGLR 23. ...... 126, 129, 131, 141
Targett v Torfaen Borough Council [1992] 1 EGLR 275, CA 31
Tarry v Ashton (1876) 1 QBD 314 248
Tennant Radiant Heat Ltd v Warrington Development Corporation [1988]
  1 EGLR 41, CA 28
Terrell v Murray (1901) 17 TLR 570 83
Terroni v Corsini [1931] 1 Ch 515 104
Thamesmead Town Ltd v Allotey [1998] 3 EGLR 97 267
Thresher v East London Waterworks Co (1824) 2 B&C 608 89
Times Fire Insurance Co v Hawke (1858) 1 F&F 406 177
Todd v Flight (1860) 9 CB (NS) 377. 246
Torrens v Walker [1906] 2 Ch 166 51
Trane (UK) Ltd v Provident Mutual Life Assurance [1995] 1 EGLR 33. ... 278
Trenberth (John) Ltd v National Westminster Bank Ltd [1980] 1 EGLR
  102 257
Tustian v Johnston [1993] 2 EGLR 8, CA 232
Twogates Properties Ltd v Birmingham Midshires Building Society
  [1997] EGCS 55, CA 123

U

Ultraworth Ltd v General Accident Fire and Life [2000] 2 EGLR
  115. .... 81, 103, 109
Unchained Growth III plc v Granby Village (Manchester) Management
  Co Ltd [2000] L&TR 186, CA 151
Underground (Civil Engineering) Ltd v London Borough of Croydon
  [1990] EGCS 40 132
Uniproducts (Manchester) Ltd v Rose Furnishings Ltd [1956] 1 WLR 45 ... 58
United Dominions Trust v Shellpoint Trustees Ltd [1993] 4 All ER
  310, CA 133
Universities Superannuation Scheme Ltd v Marks & Spencer plc [1999]
  1 EGLR 13, CA 39, 159
Upjohn v Seymour Estates Ltd [1938] 1 All ER 614. 224

V

Vaudeville Electric Cinema Ltd v Muriset [1923] 2 Ch 74 280
Vincent v Bromley London Borough Council [1994] EGCS 193, CA 48
Vural Ltd v Security Archives Ltd (1989) 60 P&CR 258 177

W

Wainwright v Leeds City Council [1984] 1 EGLR 67, CA 5, 54, 73

Wallace v *Manchester City Council* [1998] 3 EGLR 38, CA 145, 146, 148
Wallis Fashion Group Ltd v *CGU Life Assurance Ltd* [2000] 27 EG 145 64
Wandsworth London Borough Council v *Griffin* [2000] 2 EGLR
    105 .... 69, 157, 161, 162, 165
Warren v *Keen* [1954] 1 QB 15, CA 18, 19, 33
Watson v *Gray* (1880) 14 ChD 192 213
Weatherhead v *Deka New Zealand Ltd* [2000] 1 NZLR 23 .... 79, 83
Webb v *Frank Bevis Ltd* [1940] 1 All ER 247, CA 281
Wedd v *Porter* [1916] 2 KB 91 19
Weigall v *Waters* (1795) 6 Term Rep 488 26
Weinberg Estate v *Intext Linen Supply Ltd* (1992) 24 RPR 2d 75 52
Welsh v *Greenwich London Borough Council* [2000] 49 EG 118,
    CA 2, 67, 180
West Ham Central Charity Board v *East London Waterworks Co* [1900]
    1 Ch 624 17
Westacott v *Hahn* [1918] 1 KB 495, CA 50
Westminster (Duke of) v *Guild* [1983] 2 EGLR 37, CA 25
Westminster (Duke of) v *Swinton* [1948] 1 KB 524 132
Westpac Merchant Finance Ltd v *Winstone Industries Ltd* [1993] 2
    NZLR 247 38
Wheat v *E Lacon & Co Ltd* [1966] AC 552, HL 249, 250
White v *Barnet London Borough Council* (1989) 2 EGLR 31, CA 205
White v *Lord Chancellors Department* [1997] CLY 3805 251
White v *Nicholson* (1842) 4 M&G 95 33
White v *Wareing* [1992] 1 EGLR 271, CA 195
Witham v *Kershaw* (1886) 16 QBD 613, CA 20
Whitley v *Stumbles* [1930] AC 544, HL 238
Wholesale Invisible Mending Co v *Needle* [1971] CLY 6557 112
Whyte v *Redland Aggregates Ltd* [1998] CLY 3989 253
Wilchick v *Marks and Silverstone* [1934] 2 KB 56 248
Willowgreen Ltd v *Smithers* [1994] 1 EGLR 107, CA 125
Wilson v *Finch Hatton* (1877) 2 ExD 336 32
Wilson v *Stone* [1998] 2 EGLR 155 167
Wimbledon Park Golf Club Ltd v *Imperial Insurance Co Ltd* (1902) 18
    TLR 815 178
Windever v *Liverpool City Council* [1994] CLY 2816 179
Windsor and Maidenhead Royal Borough Council v *Secretary of State for*
    the Environment [1988] 2 PLR 17 274
Witham v *Kershaw* (1886) 16 QBD 613, CA 20
Wivenhoe Port Ltd v *Colchester Borough Council* [1985] JPL 396 209
Woodhouse v *Consolidated Property Corporation Ltd* [1993] 1 EGLR
    174, CA 226
Woodward v *Docherty* [1974] 1 WLR 966, CA 46
Woolworth (FW) & Co Ltd v *Lambert* [1937] Ch 37, CA 94
Wright v *Greenwich London Borough Council* [1996] CLY 4474 251

Table of cases xxxi

Wright v *Lawson* (1903) 19 TLR 510, CA 82
Wringe v *Cohen* [1940] 1 KB 229, CA 245
Wycombe Area Health Authority v *Barnett* [1982] 2 EGLR 35, CA.... 33, 186
Wykes v *Davis* [1975] QB 843, CA 235

Y
Yellowly v *Gower* (1855) 11 Exd 274 19
Yorkbrook Investments Ltd v *Batten* [1985] 2 EGLR 100, CA 50
Young v *Dalgety plc* [1987] 1 EGLR 116, CA 272

French "jurisprudence" or cases:
Cass Civ 14.10.1965, D. 1966.41 42
Civ 3e 10.12.1980, Gaz Pal 1981. 1. pan 122 43
Civ 3e 21.11.1990. Rev Dr Immob. 1991.264 43
Civ 3e 10.5.1991, D. 1991. somm.353n. 42
Civ. 3e 10.5.1991, Rev Dr Immob. 1992.124 42
Civ. 3e 13.7.1994, D. 1994.IR200 42
Civ 3e 30.11.1994, Loy et Corpr 1995 No 162 42
Civ 3e 7.1.1998, Rev Dr Immob. 1998.302 41
Montpellier 4.6.1957, D. 1957 somm.69 41
Paris 10.1.1991, D. 1991.IR50 42
Paris 15.12.1994 (cited Dalloz, Code Civil, 1997/98 n to art 1720)...... 41
Soc 29.3.1957, Gaz Pal 1957.2.59 42
Versailles 14.12.1989, D. 1990. somm.259 43

Table of statutes

Access to Neighouring Land Act 1992 . . . . 189, 257, 258, 259, 261, 267
   s1 258
      (1) 260
      (2) 260
      (3)(a) 261
         (b) 261
      (4) 260, 261, 262
         (a) 262
      (5) 261, 262
   s2 (1)(a) 262
         (b) 262
         (c) 262
      (2) 263
      (3)(a) 263
      (5) 263
      (7) 263
   s3 (1) 264
      (2)(a) 264
         (b) 264
      (3) 264
   s4 (1) 263
      (b) 263
      (4) 259
   s5 263
   s6 (1) 259
   s7 (2) 260
Agricultural Holdings Act 1984
   Sched 3, para 13 238
Agricultural Holdings Act 1986 . . . . 138, 230, 231, 232, 236, 241
   s7 (3) 231
   s8 (2) 231
      (3) 231
      (6) 231
   s22(1) 239
   s28 235
   s71 . . . . 236, 237, 238, 239, 240
      (1) 236, 237, 238, 239

Table of statutes

```
        (2) ............................................. 237
        (3) .................................. 237, 238, 239
        (4)(a) .......................................... 237
            (b) .......................................... 239
        (5) ....................................... 237, 238
    s72 ..................................... 236, 239, 240
        (1) ............................................. 240
        (2) ............................................. 240
        (3) ............................................. 240
        (4) ............................................. 239
    s73 ................................................. 236
    s74 ................................................. 236
    s83(2) .............................................. 239
    s84 ................................................. 236
    s96(1) .............................................. 231
        (3) ............................................. 237
    Sched 3, Part I ..................................... 235
            III, para 9 ................................. 235
Agricultural Tenancies Act 1995 ............... 130, 230, 241
Agriculture Act 1947
    s10(3) .............................................. 237
    s11(3) .............................................. 237
Arbitration Act 1979 .................................... 227

Building Act 1984
    s77 ............................................ 228, 229
        (1) ............................................. 228
        (2) ............................................. 228
    s78 ............................................ 228, 229
        (2) ............................................. 228
        (3) ............................................. 228
        (5) ............................................. 228
    s79 ................................................. 228

Contracts (Rights of Third Parties) Act 1999 ............. 49
    s1 ................................................... 75
        (1) .............................................. 49
        (2) .............................................. 49
Criminal Law Act 1977
    s6 (1) .............................................. 120

Defective Premises Act 1972 ............... 191, 192, 193, 194
    s1 ................................................... 30
        (5) .............................................. 30
    s4 ................................. 180, 190, 191, 192, 193
```

(1)	26, 191
(2)	192
(3)	193
(4)	192, 193
s6 (1)	191
(2)	193
(3)	193

Disability Discrimination Act 1995 .... 92, 93
 Part III 92
 s21 (2) 92
 s27 92
  (2)(6) 92
 Sched 4
  para 5 92
  para 6 93
  para 8 93

Environmental Protection Act 1990 .... 206, 208, 209
 Part IIA 55
 s78E 56
 s78F(4) 56
  (5) 56
 s79 (1) 207, 209
  (a) 207
  (7) 208
  (8) 207
 s80 (1) 207
  (2) 207, 208
  (3) 210
  (5) 211
  (6) 211
  (7) 210
 ss81–81B 207
 s81 (1) 210
 s82 211

Factories Act 1961
 s169 91
Fire Precautions Act 1971
 s28 91
Fires Prevention (Metropolis) Act 1774 178

Highways Act 1980
 s165(1) 268

Table of statutes

Housing Act 1961
   ss31–32 6
Housing Act 1980
   ss81–83 91
Housing Act 1985 205
   Part IV 197
   Part VI 202
   s96 198
   s189 205
      (1) 202
      (1A) 203
      (2) 204
      (4) 205
   s190(1)(a) 205
      (b) 205
   s193 205
   s194 205
   s207 204
   s604 .... 35, 203
      (1A) 202
   s605 202
   s610 91
   Sched 10 205
Housing Act 1988 .... 148, 181
   Part I 29
   s13 196
   s14(2)(b) 197
      (c) 197
   s16 58
   s19A 194, 196
   Sched 2A, para 6 194
Housing Act 1996 196
   Part III 155
   s81 .... 39, 119
   s82 39
   s84 168
   Sched 4 168
Housing Construction Grants and Repairs Act 1996
   s86 203
Human Rights Act 1998 .... 121, 282

Insolvency Act 1986
   s11(1)(c) 120
   s11(3)(ba) 120

s252(1) 120
Insolvency Act 2000
   s9 (3) 120

Landlord and Tenant Act 1927 . . . . 94, 112, 113, 230, 269
   Part I 93, 96, 269
   s1 (1) 270
      (2) 270
   s2 (1) 269
        (b) 96
        (c) 269
        (3) 270
   s3 (1) 269, 270
      (6) 269, 270
   s18 (1) 16, 99, 103, 104, 106, 108, 110, 111, 113, 114, 230
      (2) 125
   s19 (2) 94
Landlord and Tenant Act 1954
   Part II 27, 97, 108, 183, 201
   s24 63
   s34 . . . . 97, 108, 201
   s35 201
   s53 95
Landlord and Tenant Act 1985 . . . . 166, 182, 183, 184, 193
   s8 . . . . 29, 36, 180, 184, 185
   s11 . . . . . 1, 4, 6, 9, 25, 35, 38, 70, 87, 153, 183, 188, 189, 191, 195, 196, 197
      (1) 187
        (a) 184
        (b) 186
        (c) 186
      (1A) (a) 188
          (b) 186, 187
      (1B) 186, 188
      (2) 186
      (3) 184
      (3A) 189
      (4) 183
      (5) 183
      (6) 58, 189
   ss11–16 181
   s12 183
   s13 (2) (b) 182
        (c) 182
   s14 (4) 183
   s16(b) 183

s17	153
(2)(d)	153
ss18–30	164, 165
s18(1)	165, 172, 238
s19	46
(1)	165
(2)	166
(2A)	166
(2B)	166
s20	46, 169
(4)	166, 167
(5)	166
(a)	167
(b)	167
(c)	167
(d)	167
(g)	168
(9)	167
(B)	165
s21	168
s28	168
s29(2)	166
s31A(6)	166
s32(2)	183
s38(2)(a)	183
Sched	172

Landlord and Tenant Act 1987

Part II	25, 155
Part III	156
Part IV	168, 173
s24	155
(1)	155
(ab)	164
(b)	164
(2A)	164
(6)	155
(9A)	155
(11)	155
ss35–39	168
ss35–40	173
s40	169
s42	163
s46(1)	168
s47	168
(1)	168

| (2) 168
s48 168
Landlord and Tenant Act 1988
 s1 95
Landlord and Tenant (Covenants) Act 1995 . . . . 63, 64
 s3 . . . . 48, 49, 149
 s5 64
 s17 . . . . 102, 103
     (6)(c) 102
 s25 64
 s28(1) 48
Landlord and Tenant (Requisitioned Land) Act 1944
 s1 111
Land Registration Act 1925
 s49 263
 s54 263
Law of Property Act 1925 . . . . 117, 141
 s38 212
 s62(1) 268
 s84 91
 s146 . . . . 117, 125, 136, 139
     (1) 100, 125, 126, 127, 128, 131, 138, 140, 141
     (2) 118, 130, 135
     (3) 135, 136
     (4) 133, 134, 135
     (5)(b) 135
 s147 85
     (1) 137
     (2) 137
     (4) 137
 s196 139
     (1) 125
     (2) 125
Leasehold Property (Repairs) Act 1938 . . . . 59, 99, 100, 101, 115, 117,
 . . . . 118, 126, 136, 137, 138, 139, 140, 141, 142, 143
 s1 115
     (1) 138, 140
     (2) 130
     (3) 140
     (4) 139, 140
     (5) 115, 129, 140, 141, 142
         (a) 142
     (6) 141
 s2 136
 s3 138

Table of statutes

s6 (1) 140
s7 (1) 138
    (2) 139
Leasehold Reform Act 1967 200
    s15 (7) 200
    s20 200
Leasehold Reform, Housing and Urban Development Act 1993 200
    s36 200
    s96 (3)(a) 198
    Sched 9 200
Limitation Act 1980 217
    s32 (1)(c) 158
Local Government and Housing Act 1989
    Sched 10 194
London Building Act 1894 217
    s88 (1) 222
London Building Act 1930
    s5 216
London Building Acts (Amendment) Act 1939 .... 212, 221, 226
    s46 218
    Part VII 228

Occupiers' Liability Act 1957 .... 194, 249, 250
    s2 (1) 254
       (2) 250
       (3) 252
            (a) 252
            (b) 252
       (4) 252
            (a) 253
            (b) 253
       (5) 251
       (6) 250
    s3 251
       (1) 251
       (2) 251
Occupiers' Liability Act 1984
    s1 (3) 255
       (4) 256
       (5) 256
       (6) 256
Offices, Shops and Railway Premises Act 1963
    s73 91

Party Walls etc Act 1996 .... 212, 213, 214, 215, 216, 221, 222, 226, 227, 259

s1	224, 225
(3)	225
(4)	224, 225
(5)	224
(6)	224
s2	217, 218, 220, 227
(1)	220
(2)	220
(a)	222
(b)	213, 214, 222
(c)	223
(d)	223
(e)	223
(f)	223
(g)	223
(h)	223
(j)	223
(k)	224
(l)	224
(m)	218, 220, 224
(3)	222
(b)	222
(4)	223
(5)	223
(6)	223
(7)	224
(8)	223
s3 (1)	218
(2)(a)	219
(b)	219
s4 (1)(a)	219
(b)(i)	220
(ii)	220
(2)(a)	219
(b)	219
(3)	220
s5	219, 220
s6	220
s7 (2)	221
(5)	221
s8 (1)	221
(2)	221
s9	217
s10	218, 220, 225, 226
(1)	225

Table of statutes xli

```
        (2) ................................................. 225
        (3) ................................................. 225
        (4) ................................................. 225
        (6) ................................................. 225
        (7) ................................................. 225
        (8) ................................................. 225
        (9) ................................................. 225
        (10) ................................................ 225
        (11). ............................................... 225
        (12) ................................................ 225
        (16) ................................................ 226
        (17) ................................................ 226
    s11 (4) ............................................ 214, 222
        (5). .......................................... 213, 214, 222
        (7) ................................................. 220
    s12 .................................................... 220
    s15 .................................................... 218
        (1) ................................................. 218
    s16 .................................................... 221
    s20 .......................................... 213, 215, 216
        (1)(c) .............................................. 216
Planning (Listed Buildings and Conservation Areas) Act 1990 ....... 273
    s7 ..................................................... 274
    s9 ..................................................... 274
        (3) ................................................. 274
        (4) ................................................. 274
    s38 .................................................... 274
        (1) ................................................. 275
        (3) ................................................. 275
        (4) ................................................. 275
    s39 .................................................... 275
    s47 (1) ................................................ 276
    s48 ............................................... 276, 277
        (3) ................................................. 276
    s54 .................................................... 276
        (1) ................................................. 276
        (4) ................................................. 276
        (5) ................................................. 276
        (6) ................................................. 276
    s55 .................................................... 275
Protection from Eviction Act 1977
    s2 ..................................................... 120

Railways Clauses Consolidation Act 1845
    s68 .................................................... 268
```

Rent Act 1977 .... 80, 91, 111, 148, 181
    s3 (1) 195
       (2) 195
    s67 (3) 195
    s70 (1)(a) 196
       (3)(b) 196
    s148 58

Statute of Marlborough 1267 19
Supreme Court Act 1981
    s37 154

Theft Act 1968
    s4 (2) 15

Unfair Contract Terms Act 1977 254
    s1 (3) 254
    s2 (1) 254
       (2) 254
    s3 151

Foreign Legislation:

France
Civil Code, art 1719 41
Civil Code, art 1720 41
Civil Code, art 1721 42
Loi of 6 July 1989, art 6. 12
Decree of 6 March 1987 12

Queensland, Australia
Property Act 1974. 257
    s180(1). 257
       (2)(b) 258
       (3)(a) 258
         (c) 258

United States of America
California Civil Code, arts 1941 and 1942 12

Chapter 1

Introduction

I – Overall review of aspects of the current law

This book examines the English and Welsh legal principles which apply when property falls into disrepair. The importance of these is considerable. The government, in connection with housing policy, has admitted that: "a sizeable minority of people face severe problems with housing".[1] Some of the best known problems have been highlighted in the residential sector. The Law Commission, in a fairly recent and comprehensive report,[2] cited two cases concerned with council properties whose results were unfortunate. In one,[3] the tenant's house was affected by severe condensation dampness. The council landlord was not liable under the statutorily implied covenant applying to short residential tenancies[4] to cure this state of affairs because the property was not in disrepair. The house, as designed, was unfit for the demands of modern living which had increased the risks of the release of condensation-producing humidity.

The other case[5] concerned the tenant of a council flat which was cockroach-infested. This state of affairs lasted for some five years. Though the local authority required the landlords to abate it as amounting to a "statutory nuisance", the tenant was unable to recover any damages for her losses and inconvenience, a sum estimated at £10,000. The premises were not in disrepair.

A third case[6] further illustrates possible defects in the law. A council tenant whose flat suffered from serious condensation was held unable to bring an action against his landlord in tort,

---

[1] "Quality and Choice: A Decent Home for All" (Green Paper, 2000) p 8.
[2] Law Com No 238 (1996) *Responsibility for the State and Condition of Property* para 1.1.–1.2.
[3] *Quick v Taff-Ely Borough Council* [1985] 2 EGLR 50, CA.
[4] Landlord and Tenant Act 1985, s 11.
[5] *Habinteg Housing Association v James* (1995) 27 HLR 299.
[6] *McNerney v Lambeth London Borough Council* [1989] 1 EGLR 81, CA.

notwithstanding that the insulation and ventilation of this flat was inadequate to modern living requirements. This was owing to the fact that there is no rule in tort against a landlord letting a house in a state unfit for human habitation.[7] The landlord did not design or build the flat – so closing off that ground of action for the tenant.[8] The premises were not in disrepair, so precluding any action by the tenant in respect of the implied statutory obligation imposed on landlords.[9]

The way repairs and maintenance of commercial premises are resolved by the law, although much less controversial than in the residential sector, is not free from difficulty. In relation to a business lease, freedom of contract prevails.[10] The wording of individual repairing obligations is vital. This point is further considered later in this Chapter.

II – Some statistics

The statistical background against which the law operates is a good indication of the context of the rules governing dilapidations.[11] During the period 1970–1974, a survey was carried out of a sample size of 510 defective buildings by the Building Research Establishment.[12] It examined a selection of residential premises

---

[7] *Cavalier* v *Pope* [1906] AC 428, HL.
[8] There is always the possibility of an action against the landlord for having created a "statutory nuisance": see eg *Birmingham City Council* v *Oakley* The Times, January 8 1999 (overturned on appeal by the House of Lords, The Times, November 29 2000), but there are substantial limits to this pathway: see further Chapter 10.
[9] However, the express wording of a landlord covenant in eg a council tenancy may go further than legislation, as in *Welsh* v *Greenwich London Borough Council* [2000] 49 EG 118 (where the landlords expressly undertook to "maintain the dwelling-house in *good condition* and repair", thus requiring them to cure condensation and severe black spot in the premises, as each word of the obligation was given its proper meaning – emphasis supplied).
[10] *Marfield Properties Ltd* v *Secretary of State for the Environment* 1996 SLT 1244 (no doubt the same principle would apply in England and Wales).
[11] See further Luba and Knafler, *Repairs – Tenants' Rights* 3rd edn (1999) Chapters 1 and 2; Oxley and Gobert, *Dampness in Buildings* 2nd edn (1994); IH Seeley, *Building Surveys: Reports and Dilapidations* (1985).
[12] BRE Paper 30/75 "Building Failure Patterns and their Implications" (IL Freeman).

Introduction 3

such as council houses, council flats, private houses and also premises such as offices, schools, universities and churches. It found that among the commonest defects in housing were condensation, rain penetration and cracking. Rain penetration, for example, affected some 38% of council flats and 33% of private houses. Condensation affected 59% of council houses but only 18% of private houses.[13] About 58% of all defects were ascribed to faulty design, such as a failure to follow the criteria laid down in the relevant building regulations. A further 35% of defects were blamed on faulty execution, whereas only 12% were attributed to faulty materials and the like.[14]

A more recent survey of common defects in low-rise traditional housing[15] disclosed a list of defects such as insufficient wall ties, poor quality of mortars or bricks in the substructure, lack of damp proofing, faults in air-bricks, defective pipes and gutters, as well as insufficiently deep foundations for porches as opposed to the main house. Once again, many of these defects were put down to failure to comply with building regulations. A further survey revealed that at that time an estimated two million dwellings in England, mainly in the public and private rented sectors, might have serious damp problems.[16]

III – The specific problem of dampness

The English Housing Survey of 1996[17] confirms that dampness is and remains one of the most common reasons for housing

---

[13] BRE 30/75, *supra*, Table III. According to Seeley p 56, 29% of offices were, at the time he wrote, affected by rain penetration and 29% by cracking; 64% of factories suffered from rain penetration.
[14] BRE Paper 30/75, *supra*, Table IX.
[15] BRE Digest 268 (1988).
[16] BRE "Dampness: One Week's Complaints in Five Local Authorities in England and Wales" (1982) (Sanders and Cornish). The authors also believed that as many premises again as the two million cited experienced some troubles (p 8). They thought that dwellings in the sample from the private sector were more prone to severe damp than in the public sector (28% as opposed to 17%).
[17] DETR 1996, Chapter 6, p 59. Thus, 300,000 dwellings suffered from dampness (para 6.3). Dampness and condensation is at its most prevalent in pre-1919 dwellings whether held from local authorities or private landlords. The worst problems are concentrated in households

unfitness.[18] Damp[19] which is not cured may have a number of nefarious results. Thus, it can lead to structural deterioration; it will result in the decay of wood and will spoil decorations. It can also be dangerous to health owing to mould growth. Dampness may, in particular, manifest itself as rising damp or be the result of condensation. While a freehold owner can be compelled to remedy a damp condition if it amounts to a "statutory nuisance", it must be such as to injure the health of an occupying tenant. Moreover, enforcement depends on the discretion of a local authority, which may decide not to prosecute the owner in question.

If the damp premises are held on a lease or tenancy, one or other party to the lease may be under a covenant to keep the premises in repair. If the tenancy is for a term of less than seven years and the property is a dwelling-house or flat, the landlord is under a statute-implied covenant to keep the structure and exterior of the dwelling-house in repair.[20] However, since liability to carry out remedial work is dependent on a number of factors, notably that the party seeking to render the other liable must prove that the premises are out of repair,[21] there is no guarantee that dampness will be remedied under a repairing obligation. If the dampness is the result of condensation caused by modern living styles and there is no item in the premises which is physically out of repair or damaged as a result of the condensation whose replacement would be required under a repairing covenant, a landlord, as already seen, cannot be compelled to cure the fact that it may be owing to the inadequacy of the design of the premises for modern living that condensation has become a problem. If the cause of the dampness

---

[17] cont.
with a low level of income and poor employment prospects (see Chapter 7 of the Survey). For an analysis of the physical causes of dampness see Marshall, Worthing and Heath, *Understanding Housing Defects* (1998) Chapters 13 and 14.

[18] With, according to Sanders and Cornish, *supra*, an increased likelihood of dampness in older dwellings, since these often have solid walls and poor insulation.

[19] Defined by Oxley and Gobert as an atmosphere wetter than 85% relative humidity: a material is damp if it is in equilibrium with this humidity (p 16).

[20] By section 11 of the Landlord and Tenant Act 1985.

[21] See in particular *Post Office* v *Aquarius Properties Ltd* [1987] 1 EGLR 40, CA.

is found to be the absence of a damp-proof course in the premises, as where they were built at a time before such treatment came into being, owing to the fact that a covenant to repair does not require a landlord to improve the design of the demised premises beyond their original construction, the landlord is not liable to insert a new damp-proof course, even if thereby a problem of rising damp would be cured.[22] Since, however, a covenant to repair requires the party liable to make good a damaged item so that, after the remedial work, the item is in as good as condition as it would have been if it had not been physically damaged,[23] the landlord of flats whose damp-proof course had failed had to replace it with a new and better-designed damp-proof course than the item which had failed – even though some minor structural work would be involved.[24]

The traditional approach of English law to promoting property maintenance in the case of landlord and tenant is by means of express repairing covenants. Sometimes the lease or tenancy is silent as to repairs. In that case, implied obligations need to be considered.

IV – Implied covenant route

The courts are reluctant to imply any covenants into leases. This is largely owing to the survival of the *caveat emptor* principle – which is that the tenant takes the premises as he or she finds them – a result leading to what Lord Hoffmann characterised as the "bleak laissez faire" of the common law.[25] If parties wish to provide for the carrying out of repairs to demised premises, they must do so by suitable terms. The court will not insert any bargain they have not made into a lease.[26] If a lease or tenancy agreement would be unworkable without an implied repairing obligation by the landlord or tenant, the court is able under the guise of "business efficacy" to imply the minimum obligation needed to make the lease work.

---

[22] *Pembery v Lamdin* [1940] 2 All ER 434, CA; *Wainwright v Leeds City Council* [1984] 1 EGLR 67, CA.
[23] *Calthorpe v McOscar* [1924] 1 KB 716, CA.
[24] *Elmcroft Developments Ltd v Tankersley-Sawyer* [1984] 1 EGLR 47, CA.
[25] *Southwark London Borough Council v Mills* [1999] 3 EGLR 35 at 36L.
[26] *Hart v Windsor* (1844) 12 M&W 68 at 88 (Parke B).

Thus, a weekly statutory tenancy, entered into prior to the commencement of legislation[27] imposing specific repairing obligations on landlords, cast on the tenant an express obligation to keep the interior of the house concerned in repair but no correlative obligation on the either party to keep the exterior in repair. The landlord was held to be the appropriate person, as having more resources than the tenant, on whom to impose the latter obligation by implication, otherwise the tenant could not comply with her obligation, given the damp state of the property.[28] The narrowness of the power of the courts to imply repairing obligations is shown by the fact that the occasions where an implied repairing obligation has been imposed on landlords where a tenancy agreement is silent are almost all confined to residential tenancies.[29]

The fact that a landlord is under a statutorily implied covenant in the case of short tenancies to keep the structure and exterior of the "dwelling-house" concerned in repair[30] has, in the words of Lord Hoffmann, "inhibited the courts from developing the common law in this area".[31] The House of Lords took the view that the best place for value judgments to be made about who is to carry out repairs or cure unfitness was Parliament. The House of Lords recently refused[32] to allow two tenants of council premises which lacked sound-proofing to the standards required by current Building Regulations to bring an action under the implied covenant for quiet enjoyment against their landlords. The tenants claimed, for example, that they could hear all the noises of domestic living coming from neighbouring premises let by the landlord to other tenants. The statutory repairing covenant did not apply. To insert the latest type of sound-proofing would improve the premises, which, when built or converted, complied with the then building regulation standards. The House of Lords refused to extend the implied covenant for quiet enjoyment beyond its traditional boundaries – it was a covenant that the tenant's lawful possession

---

[27] Housing Act 1961, ss 31 and 32, now Landlord and Tenant Act 1985, s 11.
[28] *Barrett* v *Lounova (1982) Ltd* [1988] 2 EGLR 54, CA.
[29] As shown by the approach of the Court of Appeal in *Demetriou* v *Poolaction Ltd* [1991] 1 EGLR 100, refusing to imply any repairing obligation against a commercial landlord.
[30] Landlord and Tenant Act 1985, s 11.
[31] *Southwark London Borough Council* v *Mills* [1999] 3 EGLR 35 at 36M.
[32] In *Southwark London Borough Council* v *Mills*, *supra*.

of the land would not be substantially interfered with by the landlord and no more. It was "fundamental" or "settled law" that the parties were taken to intend that the landlord gave no implied warranty as to the condition of the premises.

V – Interpreting express repairing obligations

In the commercial and long residential sectors alike, where the landlord is under an express repairing obligation, as where he has let a building consisting of flats or offices to different tenants and they pay him service charges to cover the cost of the maintenance works, or where there is a lease of a single unit to an individual tenant who has covenanted to keep the premises in a specified state of repair such as in "good" repair, questions arise as to the way the courts interpret the obligation concerned. The first thing which will guide them is the actual words used by the parties in the lease, taken in its context. Each word of the particular covenant will be given its proper significance. Thus, where a landlord expressly undertook to "maintain repair amend renew ... and otherwise keep in good and tenantable condition" certain structural and exterior parts of newly-built premises, he was liable to remedy defects in the exterior cladding which were letting in water.[33] The landlords argued that an obligation to cure original defects in the construction of the building was not one which a landlord might reasonably be expected to take. The High Court, finding damage to the premises, insisted that these clear words of obligation must be given effect to. There might have been a "torrential style of drafting" in this case, but as Lindsay J pointed out[34] "even where there is a torrent, each stream of which it is comprised can be expected to have added to the flow".

A further example of the approach to the construction of covenants to repair and maintain is that of a case[35] in which the landlord was entitled to recover costs from office lessees on account not only of repairs to the premises, but also for any renewal or replacement, whenever necessary. The words of the covenant were

---

[33] *Credit Suisse* v *Beegas Nominees Ltd* [1994] 1 EGLR 76.
[34] *Ibid* at 86F. The expression "torrential style of drafting" derives from *Norwich Union Life Assurance Society* v *British Railways Board* [1987] 2 EGLR 137 at 138D.
[35] *New England Properties Ltd* v *Portsmouth New Shops* [1993] 1 EGLR 84.

perfectly clear. The High Court held that they involved both a covenant to pay repairs and for renewal. The work in this case was substantial: it involved replacing the roof of the premises (which had been seriously damaged in a storm) with a new roof which had a different pitch, and other structural improvements, at a cost of just over £202,000. The new roof was less susceptible to "wind lift".

VI – Reform of the law

The cases involving council tenants mentioned earlier suggest that some aspects of the law are in need of reform. As between commercial landlords and tenants who may be presumed to have equal bargaining strength, there is much to be said for the question of repairs and maintenance of the demised premises being regulated by their own free choice. Business people are used to making what bargain they can in accordance with their actual intentions – and so, it could be said, they should be left in a free market to do just that. Even in the commercial sector, some might question this *laissez-faire* approach. The assumption that, say, a small business person has equal bargaining strength with a large landlord company seems tenuous.[36] Nor may it always be appropriate for the law to regulate the maintenance of leasehold premises solely by reference to assumptions about parties' intentions, actual or presumed.

The Law Commission, running against the traditional freedom of contract approach, have asserted an "overriding policy consideration that it is in the public interest that the stock of leasehold property should be properly maintained".[37] In an earlier Consultation Paper[38] they had suggested that the parties to a lease should be subject to a duty to maintain the premises for the purpose for which they were intended by the tenant to be used. The duty would have referred to the safe, hygienic and satisfactory use of the property for its intended purpose. That purpose would be as stated in the lease or it would be inferred from the last use the premises had been put to before they were let. The duty would

---

[36] In the American case of *Reste Realty Corporation* v *Cooper* (1969) 33 ALR 3d 1341 at 1348 it was thought that such a person might prefer to take a lease of premises in a fit state for use. The contrary view was expressed in eg *Service Oil Co Inc* v *White* (1973) 242 P2d 652 at 659.
[37] Law Com No 238 (1996) *Responsibility for State and Condition of Property*, para 7.4.
[38] Law Commission Consultation Paper No 127 (1992), paras 5.5–5.10.

extend to the remedying of inherent defects and the making of any necessary improvements.

This radical proposal was not, it seems, well received by commercial landlords. There might not have been any advantage in a new tenant intent on using the premises for, say, a computer centre, to have let to him premises kept up in a state for a solicitor's office, if the standard for the latter was higher than the new lessee required.[39] This scheme also ran against traditional practice, whereby if a business tenant has a lease of a self-contained unit then he is, seemingly, at least where there is no glut of premises to let, ordinarily going to have to keep them in repair, and if he has a lease of a unit in a larger development, he is usually likely to have to fund a share of the cost of repairing and maintenance work in service charges payable to landlords. It could also be said that Parliament has traditionally only legislated for residential lessees, who are vulnerable, and that otherwise freedom of contract should be left, at least in the main, intact.

The Law Commission's final proposal is more limited. They recommended[40] that if and to the extent that the parties to a lease have not made provision for repairs, there should be a statutory default provision. There would be two default covenants. These could be freely excluded or modified. The first covenant would apply to the premises as let. The landlord would be subject to an implied covenant to keep the premises in repair. The repairing obligation would mean an obligation both to repair the premises or to keep them in repair, and to make and keep them fit for human habitation. The implied covenant would apply to the whole premises let and to each and every part of them. However, the Law Commission anticipated that "in many leases, particularly of business premises", the implied covenant would be expressly excluded "almost as a matter of course".[41] Their aim in proposing a statutory default covenant was to make sure that one or other party

---

[39] See PF Smith "Repairing Obligations: A Case Against Radical Reform" [1994] Conv 186. Landlords might have been discouraged from letting and would have had every incentive where lawfully possible to contract out or shift the burden to their lessees.

[40] Law Com No 238, *supra*, paras 7.7–7.31. The default covenant would not apply to short residential tenancies, which would benefit from a revised version of s 11 of the Landlord and Tenant Act 1985.

[41] *Ibid*, para 7.9.

to a lease would be liable to repair if the lease was silent. The parties would thus be encouraged to consider the issue of repairs when negotiating a lease of the premises. In addition, a second default covenant would apply to common parts and any other property under the landlord's control which might affect the enjoyment of the premises let.

The default covenants would have limits. They would be capable of being contracted out of and would only apply to leases and tenancies granted after the commencement of any legislation needed to implement the proposals of the Law Commission. They would not apply to leases of an agricultural holding, to a farm business tenancy nor to an oral lease. The default covenant provisions are modest.[42]

It is instructive in connection with these proposals to note the position north of the border. In Scotland, there is a common law duty on a landlord to let urban subjects (ie houses, offices, shop and stores) in a fit state reasonably fit for use for the purposes of the lease. The landlord must under this implied duty or "warrandice" keep the property reasonably habitable and tenantable and in a wind- and watertight condition during the currency of the lease.[43] Scottish commercial leases may well place some or all repairing obligations onto tenants. Clear language shifting the burden from landlord to tenant is required, the question being one of construction of the lease in question.[44] If the parties to a lease have deliberately failed to address the liability of a tenant for particular kinds of repair, the court may conclude that the background common law landlord obligation applies to that extent.[45] The

---

[42] See further Bridge, "Putting it Right: The Law Commission and the Condition of Tenanted Property" [1996] Conv 342 at 346.
[43] See further, Paton and Cameron, *Landlord and Tenant* (1967) pp 130ff; McAllister, *Scottish Law of Leases* (1989) Chapter 3; Gordon, *Scottish Land Law* (1989) para 19–181 *et seq*. The common law obligations have been extended in the case of certain houses and flats.
[44] *Cantors Properties (Scotland) Ltd v Swears and Wells Ltd* 1978 SC 310 at 322; also *McCall's Entertainments (Ayr) Ltd v South Ayrshire Council (No 2)* 1998 SLT 1421, where a clause shifting liability for repairs to a tenant of dilapidated subjects did not, as a matter of construction, extend to the "extraordinary repairs" required to save the property. For an example of a wide contracting out clause see *Lowe v Quayle Munro Ltd* 1997 SLT 1168.
[45] As happened in *McCall's Entertainments (Ayr) Ltd v South Ayrshire Council (No 2), supra*.

VII – Extension of liability of residential landlords

The Law Commission have also proposed an extension to the current statutory obligation of landlords to keep the structure and exterior of dwelling-houses and flats in repair. Essentially, in tenancies granted after the commencement of any implementing legislation, the landlord of a residential tenant holding for a term of less than seven years would, in addition to his current obligations to keep in repair, be subject to a statutory obligation to see to it that the dwelling-house or flat is fit for human habitation at the commencement of the lease. The landlord would also have to keep the dwelling-house or flat fit for human habitation during the lease.[46]

The reforms proposed by the Law Commission would have some limits, derived from the current law. The landlord would not be liable to carry out any work falling within a revised version of the current obligation implied by law against a tenant to use the premises in a tenant-like manner. He would also not be liable to rebuild or reinstate the house if it was destroyed by fire, flood, tempest or other inevitable accident, nor to keep in repair any tenants' fixtures. The landlord would also not be liable if the principal cause of the unfitness was the tenant's fault (as perhaps where condensation dampness was caused by his using paraffin heaters).[47]

The future direction of the law is unclear at present. It has been said that the reform proposed by the Law Commission is "long overdue", as far as residential lettings are concerned.[48] It is clear from the statistics mentioned earlier in this Chapter that unfitness continues to affect many tenants, especially those in the local

---

[46] Law Com No 238 (1996), paras 8.35–8.37.
[47] The Law Commission's example, para 8.38, is of a tenant keeping a large number of animals on the premises in breach of covenant. The landlord would be able to ask the county court, as at present, to exclude the statutory covenants in whole or in part. If the house could not be made fit for human habitation at reasonable expense, the landlord would not be liable under the implied covenant proposed: para 8.39 of the Report.
[48] Bridge, *supra* at 344; also, in stronger terms, *New Law Journal* March 22 1996, editorial.

authority and private rented sectors. The law of England and Wales would be aligned, if reformed as suggested by the Law Commission, in policy terms with that of the United States of America, where residential tenants of many states benefit from a judicial or statutory warranty of habitability.[49] One basis of this is that a short-term tenant is considered to be a product consumer, with the expectation that the landlord as hirer of the house or flat will provide him with safe and fit premises to live in. He obtains on this basis a package of services in return for his rent – thus, walls and ceilings and also adequate heat, light and ventilation, serviceable plumbing facilities, secure doors, proper sanitation and proper maintenance.[50]

The British government seems to have decided that financial constraints require the enactment of only modest reforms. In the Housing Green Paper, they express a wish to overhaul the present "pass or fail" fitness standard. They propose to introduce a health and safety rating scale, claiming that such a system would be based directly on actual hazards threatening occupants.[51] If this proposal goes ahead, the government, in its narrow reform package, will be abandoning the notion of overall housing fitness standards as seen since 1885.[52] In the same spirit, they would not implement the Law Commission proposals without "some modification",[53] the terms of which remain unknown. The reluctance of the British government to implement in full the Law Commission reforms proposed for the residential sector is regrettable. There are, however, signs across the Channel of some narrowing of tenant expectations and remedies in the residential field.[54]

---

[49] See PF Smith "A Case for Abrogation" [1998] Conv 189. The California Civil Code, arts 1941 and 1942, provides for a repair and deduct scheme for residential tenants.

[50] *Javins v First National Realty Corpn* (1970) 428 F 2d 1071 at 1074.

[51] "Quality and Choice: A Decent Home for All", (Green Paper, 2000), para 5.28.

[52] The history of unfitness legislation is traced by the Law Commission in their Report of 1996 at para 4.7–4.10.

[53] *Ibid*, para 5.29.

[54] In France, whereas a landlord of residential premises let under the 1989 regime must in principle deliver to the tenant the premises in a state fit for use and in good repair (art 6 of *loi* of 6 July 1989) he is bound, it seems, only to ensure that the premises comply with minimum norms of habitability laid down in a decree of March 6 1987 in certain cases: see Azéma, *Baux d'Habitation* (1996) p 25.

VIII – Commonholds

The burden of positive covenants such as to repair or maintain cannot run with freehold land against any successor in title to the burdened land.[55] Difficulties have therefore existed in English law in developing freehold flats, because the only reliable way positive covenants can be enforced against third parties is by means of a long lease.[56] Complaints about the management of long leasehold flats led to the setting up of a semi-official committee.[57] It reported adversely on the fact that, for example, some landlords of blocks of flats did not regularly collect service charges, and that in some developments there was no sinking fund to meet long-term or emergency repairs. There is also some pressure for freehold rather than leasehold tenure of flat units, held within a larger development, combined with a share in the common property. In contrast to a leaseholder, a freehold unit holder would have an asset which was not wasting. He might also aspire to more direct control over the management of the scheme premises than is available in the case of some long leasehold developments.

For these and other reasons explored elsewhere,[58] there is the prospect, on a time-scale dependent, no doubt, on the exigencies of Parliamentary time, of legislation to enact commonhold tenure. Until the consultation process is over, its results assimilated, and any primary legislation and relevant regulations are enacted, the exact details of any commonhold scheme remain to be seen. To achieve beneficial and non-contentious reform may be easier said than done, if only because, in contrast to Scotland,[59] England has not much background common law to build on. There have been no less than four commonhold schemes to date, the latest having been put out for consultation in late August 2000.[60] England and

---

[55] *Rhone* v *Stephens (Executrix)* [1994] 1 EGLR 181, HL.
[56] Various methods of circumventing this problem are noted in Chapter 14 of this book; also Megarry and Wade 16–019–16–026.
[57] The Nugee Committee, whose report is summarised by Hawkins [1986] Conv 12.
[58] See DN Clarke "Occupying Cheek by Jowl", Chapter 15 in *Land Law – Themes and Perspectives* (1998) ed Bright and Dewar.
[59] The Scots law of the tenement is set out in Scot Law Com No 162, *Report on the Law of the Tenement* (1998) Part 2.
[60] The latest proposals are contained in the Commonhold and Long Leasehold Draft Bill and Consultation Paper Cm 4843 (from the Lord

Wales are unusual in not having flat ownership legislation. The enactment of commonhold legislation would at least remove this anomaly.[61] At the same time, given less than encouraging indications from France,[62] the success of any commonhold legislation cannot be guaranteed. Thus problems may arise, as in some French schemes, owing to unit holder apathy in the management of the scheme property.[63] There are also thorny issues to be resolved in the event of the scheme becoming insolvent.[64]

---

[60] cont.
Chancellor's Department), the previous trilogy being "Commonhold – Freehold Flats" Cm 179 (1987); "Commonhold – A Consultation Paper" Cm 1345 (1990) and a Commonhold Draft Bill (1996). A Commonhold and Leasehold Reform Bill was published in December 2000.

[61] For an authoritative examination of the position, see Van der Merwe, "Apartment Ownership", Chapter 5 Vol VI *Property and Trust* in *International Encyclopedia of Comparative Law* (1994).

[62] See Hill "Freehold Flats in French Law" [1985] Conv 337; PF Smith "Caveat Commonholds" Chapter 8 in *Property Law: Current Issues and Debates* (1999) ed Jackson and Wilde.

[63] See also College of Estate Management Study "Is the Cure Worse than the Disease?" (1990) para 5.4.

[64] For a critique of the 1996 proposals in relation to this aspect, see Crabb "The Commonhold Association – As You Like It" [1998] Conv 283.

Chapter 2

Waste

I – Introduction

The object of waste has been stated to be to prevent a limited owner, such as a tenant for years, from despoiling the land to the prejudice of those entitled in remainder.[1] The Law Commission believe that it is rarely necessary to have recourse to an action in waste because actions for disrepair are usually brought on express or implied covenants to repair.[2] However, an action for waste is based on tort[3] and not on express or implied contract terms in a lease: it is an independent action which runs in parallel to those terms.[4] It may be brought against the tenant not only on account of his own actions but also where he is taken to be answerable for others in occupation of the land, such as his licensees.[5] An action in waste may also be of assistance to a landowner who wishes to proceed against a person who is occupying land without a formal assignment of a lease. It was also used against a director of a company, which latter occupied premises as a licensee but which was in liquidation, the tenant company itself being in financial difficulties and so presumably in no position to pay damages for the breaches of covenant which had taken place.[6] An action in

---

[1] Megarry and Wade, para 3–098.
[2] Law Com No 238 (1996) *Responsibility for State and Condition of Property*, para 10.9.
[3] With the result that the right to sue in waste cannot be assigned: *Defries v Milne* [1913] 1 Ch 98 at 109–110 (Farwell LJ).
[4] *Marker v Kenrick* (1853) 13 CB 188; *Defries v Milne, supra*.
[5] Law Comm No 238, *supra*, para 10.14. The dishonest removal by any occupier of landlords' fixtures would contravene Theft Act 1968, s 4(2).
[6] *Mancetter Developments Ltd v Garmanson Ltd* [1986] QB 1212; [1986] 1 EGLR 240, CA. The damage consisted of removing extractor fans and pipes from the external walls of the premises, used for a chemical business, without making good. These actions were procured by the director concerned and were held to be voluntary waste for which he was personally liable in damages to the owners.

waste might even be brought where a landlord has forfeited a lease, and the lessee, who has failed to obtain relief against forfeiture, damaged the premises during the obscure "twilight period" or "period of limbo"[7] between the issue of the claim form for possession and the actual date that the court orders possession to be delivered up to the landlord. Where the landlord is able to bring an action against the tenant on the covenants of the lease, some doubts have been expressed as to whether he is entitled also to bring a separate claim in waste.[8] There is, however, good authority in favour of the view that the landlord should have a choice of actions.[9] Since the landlord is not subject to the statutory ceiling on damages[10] imposed in the case of claims on the covenant to repair, he ought to be able to choose which action is most to his advantage, given that he could not recover damages both in contract and in tort.

The purpose of the tort of waste, which has been said to be a "somewhat archaic subject",[11] is to protect the landlord from permanent and serious injury to the value of his reversion. This could result from deliberate damage to the property by the tenant, or from his carrying out improvements to the property from his point of view which cause the value of the reversion to fall. The landlord is entitled to have some protection against a lessee erecting a new building on the premises which, at the end of the lease, reverts to the landlord, and the burden of repairing which will fall on him as owner. It may be that this consideration is of greatest force in the case of short leases. This may in turn explain why in some cases involving long leases, the courts have held that, assuming the lease is silent, the landlord cannot complain of the tenant making structural alterations to a building which enhance its value.[12] Thus, while the landlord is entitled to use the law of waste to protect his reversion against permanent injury to its value, he is not able to bring an action in waste on account of trifling

---

[7] *Liverpool Properties Ltd* v *Oldbridge Investments Ltd* [1985] 2 EGLR 111 at 112H (Parker LJ).
[8] In *Mancetter Developments Ltd* v *Garmanson Ltd* [1986] 1 EGLR 240 at 243 (Kerr LJ).
[9] *Kinlyside* v *Thornton* (1776) 2 Blw 1111; *Defries* v *Milne* [1913] 1 Ch 98.
[10] Landlord and Tenant Act 1927, s 18(1).
[11] *Mancetter Developments Ltd* v *Garmanson Ltd, supra* at 241; also Law Commission Report, *supra*, para 10.9.
[12] See eg *Doherty* v *Allman* (1878) 3 App Cas 709, HL.

II – Types of waste

There are three types of waste for present purposes: voluntary, permissive and ameliorating waste.

A – Voluntary waste

Voluntary waste consists in the deliberate carrying out of an act which permanently damages, injures or even destroys the land or any structures, buildings or landlords' fixtures. The removal of extractor fans and pipes from an industrial building was an act of voluntary waste, where the damage was not made good – even though the removed items were tenants' fixtures.[13] The conversion of meadow or pasture land into arable land is voluntary waste.[14] Cutting down timber also constitutes voluntary waste.[15] It must be shown that the land or buildings demised have been permanently injured. While in one case it was held that both a tenant and his sub-lessee were liable in voluntary waste for dumping material on land whose rental value was in fact increased as a result,[16] the authority relied on in support[17] was unreliable. It has arguably been superseded by the more liberal approach of the nineteenth century and later cases[18] favouring the avoiding of implied user restrictions.[19] Thus the better view is that in cases of this kind, it is a question of fact and degree whether the tenant's actions are voluntary or ameliorating waste.

The destruction of a building (at least where it is not replaced) is an act of voluntary waste but the tenant has a defence if the

---

[13] *Mancetter Developments Ltd v Garmanson Ltd, supra.*
[14] *Simmons v Norton* (1831) 7 Bing 640; *Rush v Lucas* [1910] 1 Ch 437.
[15] *Honywood v Honywood* (1874) LR 18 Eq 306. "Timber" for this purpose is oak, ash and elm trees at least 20 years old; Foa, p 286; but cutting down bushes or underwood is not waste.
[16] *West Ham Central Charity Board v East London Waterworks Co* [1900] 1 Ch 624.
[17] *Lord Darcy v Askwith* (1617) Hob 234.
[18] See the analysis in 43 (1929–30) Harv LR 1130.
[19] See Bathurst "The Strict Common Law Rules of Waste" (1949) 13 Conv (NS) 278.

building had been destroyed owing to his using it in an ordinary and reasonable manner for the purpose for which it had been let, as where the floor of a warehouse accidentally collapsed while the tenant was using it for the purposes of his business.[20]

B – Permissive waste

Liability for permissive waste is based on neglect to repair or maintain. Thus liability arises if a tenant allows buildings on the demised land to collapse for lack of repairs,[21] but not if the house was dilapidated at the commencement of the lease.[22] If a house is allowed to become uncovered at the roof, causing the timbers to rot, this is an act of permissive waste, unless at the commencement of the tenancy the house had no roof. It is easier to say what is not permissive waste than to indicate what conduct will incur liability.[23] Cases of permissive waste would not be very likely today, owing to the fact that even in the absence of express covenant to keep in repair, the tenant is subject to an implied obligation to use the premises in a tenant-like manner.[24]

C – Ameliorating waste

The main basis of liability in waste is that the landlord's reversionary interest in the land has been permanently and substantially damaged by the action of the tenant. If the tenant has no covenant against making structural or other alterations to the premises, and carries out an alteration which improves the value of the premises, there is ameliorating waste, which is not actionable. Thus, both where a long lessee converted dilapidated storage buildings into dwellings, so increasing their value,[25] and a tenant converted arable land into market gardens,[26] the landlords' actions failed. By contrast, it was held to be voluntary but not ameliorating waste for a yearly tenant to have rebuilt a dwelling-house and shop

---

[20] *Manchester Bonded Warehouse Co v Carr* (1880) 5 CPD 507.
[21] 2 Co Inst 145; *Herne v Bembow* (1813) 4 Taunt 764.
[22] Co Lit 53a.
[23] Law Com No 238, *supra*, para 10.20.
[24] *Warren v Keen* [1954] 1 QB 15, CA.
[25] *Doherty v Allman* (1878) 3 App Cas 709, HL.
[26] *Meux v Cobley* [1892] 2 Ch 253.

as a large shop.[27] However, the conduct of the tenants in this case amounted to a breach of their implied obligation to use the premises in a tenant-like manner. Where a long lessee carries out structural alterations, if the court finds that there was no "substantial and real"[28] damage to the landlord's reversion, it will decline both to award any substantial damages and to restrain the tenant by injunction.[29]

III – Extent of liability for waste

A tenant for a fixed term is liable for voluntary waste and also, unless the tenancy makes contrary provision, for permissive waste, as from the Statute of Marlborough 1267.[30] According to a work of authority,[31] a periodic tenant (for a yearly or for a shorter period such as for a quarter or a month) is liable for voluntary but only, in contrast to a tenant for a fixed term, for permissive waste to the extent of keeping the premises wind- and water-tight.[32] The sole liability of a weekly periodic tenant, after the re-examination of the position by the Court of Appeal,[33] is to use the premises in a tenant-like manner. Although as a rule a liability in waste co-exists with any contractual liabilities, express or implied in the lease,[34] this latter principle is one example of the tort liability having been superseded.

A tenant at will is not liable for voluntary or permissive waste, since it is generally assumed, despite doubts,[35] that an act of waste

---

[27] *Marsden v Edward Heyes Ltd* [1927] 2 KB 1, CA.
[28] Per Lord Blackburne in *Doherty v Allman, supra* at 734.
[29] *Doherty v Allman supra*, per Lord O'Hagan at p 726; see also *Doe d Grubb v Earl of Burlington* (1833) 5 B&Ad 507 (pulling down of valueless barn not waste) and *Hyman v Rose* [1912] AC 623 (making a new door during conversion of premises from a chapel to a cinema not of itself waste).
[30] *Yellowly v Gower* (1855) 11 Exd 274 at p 294 (Parke B); see now *Dayani v Bromley London Borough Council* [1999] 3 EGLR 144, where the authorities going back to the thirteenth century were considered in extenso and the contrary view of *Woodfall* 13–124, doubting liability in permissive waste, was rejected.
[31] Megarry and Wade, para 14–233. Fair wear and tear is excepted.
[32] As to which see *Wedd v Porter* [1916] 2 KB 91.
[33] In *Warren v Keen, supra*; also Chapter 3 of this book.
[34] *Marsden v Edward Heyes Ltd, supra*.
[35] *Halsbury's Laws of England*, Vol 27, para 348.

automatically terminates his interest by operation of law, and the tenant at will becomes a trespasser.[36] A tenant at sufferance is liable for voluntary waste.[37]

IV – Remedies for waste

The liability of a tenant for waste may be enforced by an action for damages. The nature of the claim is not necessarily the same as that for breach of covenant to deliver up the premises at the end of the term in the same state as the tenant received them.[38] The measure of damages is the amount of injury to the landlord's reversion. This might be the amount of the drop in its sale value, with a discount for immediate payment owing to the fact that the reversion is falling in later and not at the time of the claim.[39] A damages award may properly include the cost of making good the damage to the premises caused by the removal of tenants' fixtures without making good the damage.[40] It will be appreciated that this sum may not necessarily be equivalent to the cost of reinstatement of the premises. Even if there is a serious case of voluntary waste, exemplary damages are not appropriate since waste was not one of the exceptional classes of case reserved as permitted for punitive damages by the House of Lords.[41]

The court has a discretion to award a negative injunction to restrain an alleged act of voluntary waste. A tenant who committed permissive waste could not, it appears, be compelled by a mandatory injunction to carry out the necessary repairs.[42] In relation to ameliorating waste, where the value of the reversion was increased by the improvements of the tenant, the House of Lords refused an injunction in the exercise of its discretion.[43]

---

[36] *Countess of Shrewsbury's Case* (1600) 5 Co Rep 13b; *Harnett v Maitland* (1847) 16 M&W 257.
[37] *Burchell v Hornsby* (1808) 1 Camp 360; 170 ER 985 (a case on holding over). Megarry and Wade, para 14–236 think that such a tenant is "probably not" liable for permissive waste.
[38] Foa, p 287.
[39] *Whitham v Kershaw* (1886) 16 QBD 613, CA.
[40] *Mancetter Developments Ltd v Garmanson Ltd* [1986] 1 EGLR 240, CA.
[41] In *Rookes v Barnard* [1964] AC 1129.
[42] *Powys v Blagrave* (1854) 4 De GM&G 488; *Barnes v Dowling* (1881) 44 LT 809; Foa, p 287.
[43] *Doherty v Allman* (1878) 3 App Cas 709.

V – Reform of the law of waste

The Law Commission examined the law of waste. They identified a number of defects in it. These included doubts as to liability for permissive waste, questions as to whether the landlord may sue in waste if the lease contains an express covenant to repair, uncertainties as to the liability of a licensee in waste, and the relationship between waste and the obligation of tenant-like user.[44] With a view to providing one single cause of action applying to leases, periodic tenancies, tenancies at will, tenancies on sufferance and licences, the Commission recommended a default obligation, which the parties would be free to modify or exclude, in the form of a new implied statutory covenant. This obligation would, however, apply where the tenancy agreement or licence was informal. The main reform proposed is that a tenant or licensee would undertake to take proper care of the premises let to him or of which he was in occupation or possession. He would also be liable to make good any damage wilfully done or caused to the premises by him, or by any other person lawfully in occupation or possession of or visiting the premises. The tenant or licensee would be unable to carry out any alterations or other works whose actual or probable result would be to destroy or alter the character of the premises or any part to the detriment of the landlord or licensor.[45] The Commission say that the new covenant would capture the essence both of the law of waste and of the implied duty to use in a tenant-like manner, and to restate these in a clear form. It will be seen that the obligation is minimal. Only if there had been deliberate damage would an obligation to repair be imposed: otherwise, as at present under the implied duty to use in a tenant-like manner, the tenant or licensee is subject to a duty of routine maintenance.

---

[44] Law Com No 238, para 10.31.
[45] Law Com No 238, para 10.37. Compare the more radical proposals of the Law Commission Consultation Paper (1992), para 5.58ff.

Chapter 3

Implied obligations of landlord and tenant as to repair and fitness

I – General principles

The common law has, in Lord Hoffmann's words,[1] a "bleak laissez faire" approach to the question of implying into a lease or tenancy agreement covenants to repair or keep in a fit state for use or habitation by a landlord who has not undertaken express obligations to repair or warranties as to fitness or suitability for use. A repairing covenant cannot be implied into a lease under the guise of the "usual covenant" doctrine.[2] The refusal, save exceptionally, to imply repairing obligations or related duties, such as warranties of fitness, has been based on freedom of contract, thus expressed in an old case: "it is much better to leave the parties in every case to protect their interests themselves, by proper stipulations".[3] The courts regard leases as passing to the tenant the risks of the state of the property. Thus the tenant, if he cannot procure an express warranty from the landlord, must "satisfy himself that the premises are in good repair and fit for the purpose he wants to use them".[4]

In principle, therefore, if a landlord is to be liable to the tenant for the state and condition of the property, he must either undertake an express repairing covenant, or be subjected to a statutory obligation to keep in repair, or he must have given the tenant an express warranty as to the state and condition or the fitness and suitability of the premises for the tenant's purpose. The House of Lords have refused to extend the scope of an express or implied covenant for quiet enjoyment so as to require a council landlord to install new sound-proofing as required by the latest building regulation standards in two flats, built or converted before those regulations

---

[1] In *Southwark London Borough Council* v *Mills* [1999] 3 EGLR 35 at 36, HL.
[2] *Doe* v *Withers* (1831) 2 B&Ald 896.
[3] *Hart* v *Windsor* (1844) 12 M&W 68 at 88 (Parke B).
[4] *Edler* v *Auerbach* [1950] 1 KB 359 at 374.

came into force.⁵ The House of Lords noted that statute has modified the common law principle of no liability. They declined to modify it further, even though Lord Hoffmann admitted that a council tenant did not have the bargaining power to exact any express warranty as to the condition of the premises or the freedom of choice to reject property which did not meet his needs.⁶ The remedy for the current state of the law lay in the view of the House of Lords, with Parliament, which alone could resolve the issues of priority in the allocation of resources.

The approach of the common law has sometimes been formulated by reference to a test of business efficacy. This is a narrower principle than asking whether it is reasonable to imply the term sought: the House of Lords have refused to imply terms on the latter basis.⁷ However, it was inconsistent with the whole relationship of landlord and tenant not to imply a term in the nature of an easement allowing tenants of flats to make use of the common parts such as common stairways and rubbish chutes. Thus, the landlord had to keep the rubbish chutes, common stairways and other common parts in reasonable repair. The term was only implied as a matter of "business efficacy". It provided no guarantee by the landlords against the results of repeated vandalism.⁸

In one case, however, the Court of Appeal seemed to adopt a broader test than that of business efficacy. It implied an obligation by the landlord of a statutory periodic tenant to carry out certain specified repairs to the structure and exterior of the relevant building. The tenant was under an express obligation to carry out internal repairs. Since these could not be done because the external walls were persistently damp, the court, rejecting the view that there was no general power to imply repairing obligations into leases, imposed a correlative obligation to repair on the landlord as the party best able to shoulder that obligation.⁹ There were special facts – it may have been unusual for a statutory tenant to have any repairing obligations and for the landlord to have none. In any case, legislation first enacted some 20 years after the tenancy considered

---

⁵   *Southwark London Borough Council* v *Mills* [1999] 3 EGLR 35.
⁶   *Ibid* at 36.
⁷   *Liverpool City Council* v *Irwin* [1977] AC 239.
⁸   *Liverpool City Council* v *Irwin, supra*.
⁹   *Barrett* v *Lounova (1982) Ltd* [1988] 2 EGLR 54.

by the Court of Appeal was entered into[10] provides for repairing obligations to be implied against landlords of short residential lessees.

Where the parties to a lease of business premises have for some reason made no provisions in their contract as to the repair of the premises, the common law is unwilling to imply repairing obligations against the landlord, unless both parties would unhesitatingly have agreed that the suggested term was essential.[11] In contrast to the position in the residential sector, where Parliament has intervened, with the effect of inhibiting the development of the common law,[12] the courts, when dealing with business leases, are prepared to contemplate the fact that "there may be situations in which there is no repairing obligation imposed ... on anyone in relation to a lease".[13]

If a lease contains express terms governing repairs, even if these are not comprehensive, the courts ordinarily rule out implying repairing or fitness obligations against the landlord, if only on the basis that filling in gaps in what is, no doubt, taken as a freely negotiated contract, is not for the court. Thus, it was said that "a question of implication ultimately turns on a detailed examination of the terms of the lease".[14] The courts have refused to imply contractual terms if the remedies within the scheme of the lease are deemed adequate to meet the case.[15]

---

[10] Landlord and Tenant Act 1985, s 11.
[11] *Demetriou* v *Poolaction Ltd* [1991] 1 EGLR 100, CA.
[12] Lord Hoffmann in *Southwark London Borough Council* v *Mills* [1999] 3 EGLR 35 at 36.
[13] *Demetriou* v *Poolaction Ltd* [1991], supra at 104E (Stuart-Smith LJ). But see *Holding & Barnes plc* v *Hill House Hammond Ltd* [2000] L&TR 428.
[14] *Hafton Properties Ltd* v *Camp* [1994] 1 EGLR 67 at 70A.
[15] As in *Duke of Westminster* v *Guild* [1983] 2 EGLR 37, where the tenant was expressly entitled to enter the landlord's adjoining land if he needed, under his own repairing duties, to clear a drain running from there to his premises. Similarly, where statute provided a sufficient remedy, the High Court declined to imply any terms. In *Hafton Properties Ltd* v *Camp*, supra, the tenants of long leases of flats could pray in aid the remedies of Landlord and Tenant Act 1987 Part II. Hence, there was no necessity to imply a term that the landlord must intervene if the management company concerned failed to carry out maintenance to the premises. Besides, the leases expressly permitted (but did not require) the landlord to intervene in the event of such default.

In addition, if it would be inconsistent with the actual or presumed intentions of the parties to imply repairing obligations, the courts decline to do so. Thus a claim by a long lessee, that his lessors were liable for structural repairs caused by gradual deterioration to the building concerned, which might not be covered by an insurance policy maintained by the lessors, was dismissed as "untenable".[16] The context was examined, including the fact that the claimant lessee held a lease of some 260 years in duration and that detailed provisions were made in the original development for the carrying out and financing, by lessees, of repair and maintenance work to the premises.

II – Implied obligations of landlord to repair

A – Land, buildings or unfurnished houses

The following points show the strength of the principle that the courts will ordinarily not imply any repairing obligations against a landlord, where the tenant cannot benefit from an express covenant or warranty from him, and the narrowness of the exceptions to it at common law.

1. The landlord is under no implied duty to the tenant to put the premises into repair before letting them.[17] He is under no implied duty to keep the premises in repair during the lease.[18]
2. The fact that the landlord may have reserved himself an express right to enter and execute repairs makes no difference.[19] If the court implies a right of entry to do repairs, as against the landlord of a weekly residential tenant, this will trigger a liability to execute repairs under section 4(1) of the Defective Premises Act 1972.[20]
3. If the tenant is subject to an express duty to keep premises in repair subject to a stated excepting event, such as fire or fair wear and tear, if the event materialises, the landlord is not impliedly bound to carry out repairs or reinstatement.[21] Since

---

[16] *Adami* v *Lincoln Grange Management Ltd* [1998] 1 EGLR 58 at 60C.
[17] *Hart* v *Windsor* (1844) 12 M&W 68.
[18] *Gott* v *Gandy* (1853) 2 El&B 845.
[19] *Sleafer* v *Lambeth Borough Council* [1960] 1 QB 43, CA.
[20] *McAuley* v *Bristol City Council* [1991] 2 EGLR 64, CA.
[21] *Weigall* v *Waters* (1795) 6 Term Rep 488.

the tenant remains liable for rent, in the absence of a suspension of rent clause, a lease may expressly require the lessor to reinstate the premises in the event of fire, using insurance moneys.
4. A landlord was held liable to carry out external painting where the tenant was expressly obliged to pay for the cost of this work.[22]
5. Specific legislation enables the court to enforce landlords' repairing covenants indirectly, as by the appointment of a manager or a receiver (see Chapter 8), so militating against any implication of obligations in such cases.[23]
6. The common law has not been prepared, so far as is known, to imply any repairing obligations against the landlord of a tenant entitled to claim a renewal under Part II of the Landlord and Tenant Act 1954.

B – Offices and unfurnished flats

Where there is a lease of offices or flats in a multiple-occupied building, the lease may well provide a scheme for the carrying out of repairs by the landlord (or sometimes in the case of flats, as by means of a lessees' association) and a system for payment of the cost of works by service charges levied on individual lessees. If the lease fails to make provision for these matters, the common law only fills gaps exceptionally, on principles already mentioned. It might thus be prepared to hold that an obligation to pay service charges carries with it a correlative duty to carry out repairs. There are exceptional cases where the courts have imposed implied duties on landlords to carry out repairs, the first in tort and the second in contract.

Landlord demises only parts of premises

Should a lease expressly not demise to individual tenants of a multi-occupied building a thing such as steps or a staircase, which is used by lessees as a common means of access, the landlord is under an implied duty in tort to take care to keep the steps or staircase reasonably safe.[24] The tortious rule extends to a case where

---

[22] *Edmonton Corporation* v *WM Knowles & Son Ltd* (1961) 60 LGR 124.
[23] As in *Hafton Properties Ltd* v *Camp* [1994] 1 EGLR 67.
[24] *Dunster* v *Hollis* [1918] 2 KB 795; *Cockburn* v *Smith* [1924] 2 KB 119, CA.

the landlord retains under his control something ancillary to the demised premises, such as a roof or guttering, where the item must be kept in proper repair so as to protect the demised premises or so as to ensure that they may safely be used by the tenant. Thus, where a rain-gutter in a roof, in the landlord's control, became stopped up, the lessor's failure to remedy the known disrepair caused them to be liable for the losses occasioned to their tenants.[25] The scope of this principle is unclear, but a landlord was held liable in nuisance for having allowed rain-water outlets to certain unlet units which he controlled to become blocked, to his knowledge, and for having failed to clear these.[26] The damage to the complainant tenants was reasonably foreseeable. An implied tort duty is, however, not absolute. The landlord is bound to take reasonable care to ensure that damage does not take place to the lessee's premises, as by making satisfactory arrangements for periodical inspections.[27] He is not liable for fortuitous happenings, as where a box-gutter had been gnawed through by a rat, leading to an escape of water into the lessee's premises.[28]

III – Warranties of fitness

A – Rules applying to tenancies generally

General issues

The common law imposes no implied obligation on the landlord to guarantee that the premises when let will be physically or legally fit for the purpose for which they have been demised,[29] whether this is for habitation, business purposes or any other use. Hence, a landlord was held not liable to a tenant who took a lease of land to be used for grazing horses, when the land was poisoned by refuse heaps.[30] The House of Lords has reaffirmed the common law no-liability principle. It refused to reform it by judicial fiat. Lord

---

[25] *Hargroves, Aronson & Co v Hartopp* [1905] 1 KB 472.
[26] *Tennant Radiant Heat Ltd v Warrington Development Corporation* [1988] 1 EGLR 41, CA.
[27] As in *Kiddle v City Business Properties Ltd* [1942] 1 KB 269.
[28] *Carstairs v Taylor* (1871) LR 6 Ex 217.
[29] *Dialworth Ltd v TG Organisation (Europe) Ltd* (1996) 75 P&CR 147 at 149 (Henry LJ).
[30] *Sutton v Temple* (1843) 12 M&W 52.

Millett considered that the only proper forum for changes was Parliament. It alone was capable of making the necessary judgments as to the allocation of financial resources.[31]

Statute provides no general redress for residential tenants who wish to complain of unfitness for habitation as opposed to disrepair.[32] Section 8 of the Landlord and Tenant Act 1985 provides, it is true, that it is an implied condition of a tenancy of any length of a house for human habitation that the house is fit for human habitation at the commencement of the tenancy. It also imposes on the landlord an implied duty in the form of an undertaking to keep the premises fit for human habitation during the tenancy. However, this legislation is a dead letter. The rent levels were last fixed in 1957 at a time of rent control: they are an annual rent of £80 in London and £52 elsewhere. Judicial notice has been taken of the redundancy of these provisions,[33] even before rents were deregulated in 1988.[34]

Unfitness and disrepair

If premises are unfit for use, and the landlord is under an express or statutory covenant to repair, the latter obligation may not necessarily require him to remedy the condition complained of, owing to the distinction between disrepair and unfitness.[35] Thus, while a council house was unfit for use because its original design caused it to suffer from severe condensation under modern living conditions, the landlord was not liable under the statutory covenant to repair (which takes effect as an implied repairing obligation) to cure the unfitness of the house by replacing undamaged windows with items of a different design so as to reduce the "sweating" of the current metal frame windows. They were not physically damaged, thus not out of repair. The statutory obligation to "keep in repair" was not triggered.[36] There is force in

---

[31] *Southwark London Borough Council* v *Mills* [1999] 3 EGLR 35 at 44.
[32] Local authority powers to require the remedying of a statutory nuisance depend on the authority invoking its powers, which may in due course lead to a criminal prosecution. See further Chapter 10.
[33] *Quick* v *Taff-Ely Borough Council* [1985] 2 EGLR 50, CA.
[34] By the introduction of assured and assured shorthold tenancies by Part I of the Housing Act 1988.
[35] As pointed out by Brooke LJ in *Issa* v *Hackney London Borough Council* [1997] 1 All ER 999 at 1007–1008.
[36] *Quick* v *Taff-Ely Borough Council, supra.*

the view of the Law Commission[37] that "there is little point in making provision for repair if it does not ensure that the property is fit to live in".

Trapped in the contract

If there is not much prospect of obtaining redress for unfitness by means of a contractual implied term, the position is equally bleak for tenants in the law of tort. In general, it remains the case that, fraud apart, there is no law against letting a tumble-down house.[38] Hence, a landlord was not liable in tort to strangers to the tenancy, such as the tenant's wife, for a dangerous defect in unfurnished premises which was present when the property was let and which fell outside the covenant to repair because the tenant failed to prove that the item was in disrepair.[39]

This rule has lasting vitality, as shown by the result in a fairly recent case.[40] A flat in a block buit in the late 1940s or early 1950s to the then correct design standards had been constructed with solid walls and steel window-frames. It was let by a local authority. The flat suffered from condensation, the walls operating as "cold radiators". The insulation and ventilation of the property were deficient, having regard to modern living styles, such as not using heating all day but only at intervals, and using washing machines and tumble-dryers. No part of the flat was physically damaged and so it was not out of repair. The Court of Appeal held that the landlord was not liable in tort for the fact that the flat was unfit for use. It rejected an argument that there should be a general duty in tort to take reasonable care, imposed on landlords, to see to it that premises were habitable when let. Thus, if a tenant is unable to bring an action in contract or statute, because there is no disrepair, and the landlord has not designed or built the house, so ruling out a claim under that head, the general no-liability rule in tort will deny him a remedy.[41]

---

[37] Report of 1996, *supra*, para 5.16.
[38] *Robbins* v *Jones* (1863) 15 CB (NS) 221.
[39] *Cavalier* v *Pope* [1906] AC 428.
[40] *McNerny* v *Lambeth London Borough Council* [1989] 1 EGLR 81, CA; PF Smith [1989] Conv 216.
[41] The tenant might be able to claim under Defective Premises Act 1972, s 1, but the limitation period may preclude this – commencing six years from "completion" of the building (s 1(5)).

Exceptions to tort rule

The tort no-liability rule has developed some exceptions. If the landlord has negligently designed and built a house or flat, or some feature in it, then if he was responsible for putting in the defects complained of, he is liable in negligence for personal injuries caused to the tenant or to any persons whom the landlord might reasonably expect to use the premises. Thus, a local authority which designed and built a flat was liable to a tenant who was injured by putting his hand through a defective but pre-installed glass panel, built into an internal wall, which had a defective design.[42] Likewise, a council landlord who constructed and designed a house with inadequately lit steps with no handrail for part of their span was liable in negligence to a weekly tenant for personal injuries caused, even though the tenant was aware of the defects, as, on the facts, it was not reasonable to expect the plaintiff to avoid the risk of injury by leaving the house or himself providing a handrail.[43] By contrast, a local authority which had installed window-locks with removable keys, which were held not to be negligently designed by the standards of a reasonably skilful window designer and installer had not broken any duty of care to the tenant or other occupiers of the house.[44]

B – Lettings of furnished houses

Exceptionally, in the case of a letting of a furnished house or furnished flat, there is an implied contractual undertaking by the landlord, that at the commencement of the tenancy, the premises are in a fit state for habitation. If this undertaking, which is a condition of the tenancy, is not fulfilled on the day the tenancy commences, the tenant may repudiate the tenancy and quit,[45] without further liability for rent. The tenant may take the tenancy, and sue for damages. If he elects to repudiate the tenancy, he is not under an obligation to give the landlord an opportunity to do any

---

[42] *Rimmer* v *Liverpool City Council* [1984] 1 EGLR 23, CA.
[43] *Targett* v *Torfaen Borough Council* [1992] 1 EGLR 275, CA.
[44] *Adams* v *Rhymney Valley District Council* [2000] 39 EG 144. The council were held to have a reasonable choice between the design used and button locks.
[45] *Smith* v *Marrable* (1843) 11 M&W 5.

necessary repairs to the house.[46] It is also no defence to liability that the landlord honestly thought the house to be habitable, when let, if this is not objectively the case.[47] The policy of this rule is seemingly that where a person takes a letting of a furnished house or flat for a limited time, as for a seasonal letting, he expects to find it reasonably fit for habitation on the day he enters.[48] This implied undertaking was held to have been broken by various types of matters which would make the premises unfit for human habitation, such as defective drains,[49] as well as bug infestation.[50]

This limited exception to the common law rule of no liability does not extend beyond the state of the house at the commencement of the tenancy. The landlord is under no implied obligation that a furnished house or flat will continue to be fit for habitation during the term. If, therefore, subsequently to the start of the tenancy the premises become unfit, the tenant will not be able for that reason to repudiate the tenancy.[51]

IV – Implied obligations of tenant

The implied obligations of a tenant who is not under an express covenant to repair depend on two sets of rules. The first set is to the effect that the tenant is – no matter what the length of the tenancy may be – bound not to commit voluntary waste.

The second set of rules depend on implied covenant, which, where applicable, seems largely to make the law of waste redundant since implied covenant covers the same field. However, a tenant holding under a term certain, as opposed to a periodic tenancy, will probably have whatever repairing obligations to which he may lawfully be subjected defined expressly in his lease. The cases concerned with implied tenant covenants were concerned with periodic tenants.

Any periodic tenant is, in the absence of express covenants on his part, subject to a continuing implied undertaking to use the demised premises in a proper and tenant-like manner and to

---

[46] *Wilson v Finch Hatton* (1877) 2 ExD 336.
[47] *Charsley v Jones* (1889) 53 LP 280.
[48] *Wilson v Finch Hatton, supra* at 342.
[49] *Wilson v Finch Hatton, supra.*
[50] *Smith v Marrable, supra.*
[51] *Sarson v Roberts* [1895] 2 QB 395, CA.

deliver up possession of the premises to the landlord in the same condition as they were in at the commencement of the tenancy, fair wear and tear excepted.[52] This implied covenant will not, apparently, be displaced by an express covenant to leave the premises in a certain state of repair so that presumably the landlord may claim for breach of either covenant.[53] However, an implied covenant will be displaced by any tenants' express repairing covenant applying during the term of the lease.

The work that may be required of the tenant by this implied covenant is not particularly onerous. In the case of a weekly tenant and any periodic tenant, the principle is as follows.[54] The tenant must repair damage to the premises caused, wilfully or negligently, by him, his family and his guests. He must take proper care of the property. He must do those little jobs about the house (or flat) which a reasonable tenant would do, such as chimney-cleaning, mending the electric light when it fuses, unstopping the sink when it is blocked by his waste. If the house falls out of repair owing to fair wear and tear, lapse of time or for any reason not caused by the tenant's action or negligence, then he will not be liable to repair the damage. As a result, a weekly statutory tenant was held not liable to pay for repairs to damp and decaying walls of a house, since the disrepair was caused merely by the lapse of time which had caused the walls to require repointing. Nor was this tenant liable to pay for repairs to decayed window-sills, which had fallen into that condition for want of external repainting at regular intervals.

The operation of the tenants' implied duty is shown by two contrasting results. In one case, a tenant was not required to drain the water system and turn off the stop-cock on leaving the house in question for three nights in mid-winter. Nor was the tenant inevitably to be expected to lag water-pipes in the house to protect them against freezing or to keep the house heated as an alternative precaution.[55] The tenant thus escaped liability for damage caused to the house by a burst pipe because it was not reasonable to anticipate the pipe freezing during a short absence, given the temperature when the tenant left. It was said that the precautions which the tenant is obliged to take under the standard required by

---

[52] *Marsden v Edward Heyes Ltd* [1927] 2 KB 1, CA.
[53] *White v Nicholson* (1842) 4 M&G 95.
[54] *Warren v Keen* [1954] 1 QB 15, CA.
[55] *Wycombe Area Health Authority v Barnett* [1982] 2 EGLR 35, CA.

tenant-like user will depend on the circumstances, such as the severity of cold conditions in the house and the length of any contemplated absence. By contrast, in a Scottish case,[56] the tenant of a villa left it unoccupied for a month in mid-winter without having turned off the water or emptying the cisterns. She was held liable for damage caused when the pipes burst: she failed to use a reasonable degree of diligence to preserve the premises from injury.

V – Reform of the law

The Law Commission have made detailed recommendations in connection with implied repairing and fitness obligations.[57] "The circumstances in which a court will imply a repairing obligation are uncertain and unpredictable", it was said, and, owing to the "presumption against implying repairing obligations", the law did not encourage repair.[58] The Commission's intention was that in every lease either landlord or tenant would be responsible for the premises let, any common parts and any other premises owned by the landlord "that may impinge on the enjoyment of the property leased".[59] We here revisit the proposals affecting short residential tenancies, whose result would be to extend the reach of the current statutory rules in a little more detail than in Chapter 1 – even though, for reasons there mentioned, it is likely that these reforms will not be enacted without modifications, the shape of which remains to be seen.

The reforms as a whole rest on the premise that "tenants of residential properties under short term leases should have civil remedies against their landlords if those properties are not fit for human habitation".[60] The Law Commission decided to abandon a rent limits approach and to adopt one based on the length of the lease, as already is the case with repairing obligations. The Commission thought that the problem of unfitness was widespread enough to require some legislative action. They noted that with deregulation of rents since 1989, the fitness of such premises should be a pre-condition of the product landlords put on the rented

---

[56] *Mickel* v *M'Coard* 1913 SC 896.
[57] Law Com No 238 (1996) *Responsibility for State and Condition of Property*.
[58] Law Com No 238, para 7.3.
[59] *Ibid*, para 7.4, cf *Hallisey* v *Petmoor Developments Ltd* [2000] EGCS 124 for an express scheme.
[60] *Ibid*, para 8.10.

market.⁶¹ This useful reform package is, however, cautious. Thus, the notice rule (see Chapter 10) would not be done away with. The work of curing unfitness would only fall on a landlord who could remedy it at reasonable expense, seeing that the new obligation would not apply if the property cannot be made fit for human habitation at reasonable expense.⁶² In addition, the proposals would only apply to tenancies entered into after any new legislation came into force.

The Law Commission proposed to imply into any lease of a dwelling-house for a term of less than seven years a covenant by the lessor that the dwelling-house is fit for human habitation⁶³ at the commencement of the lease, and a covenant that the lessor will keep it fit for human habitation during the lease.⁶⁴ The repairing regime of section 11 of the Landlord and Tenant Act 1985 would also be retained.

The revised and expanded implied obligation would, with some specified exceptions,⁶⁵ apply to any lease under which a dwelling-house was wholly or mainly let for human habitation. As with the present law, this new statutory implied warranty would not apply to tenants' works required by a revised implied tenants' obligation.⁶⁶

The Law Commission thought that it was desirable in the public interest that the landlord should correct inherent defects to the property.⁶⁷ Although unfitness may be dealt with currently under local authority statutory powers, these are discretionary. The statutory nuisance provisions, as the Law Commission pointed out,⁶⁸ are aimed at dealing with the nuisance, not with providing the tenant with remedies such as damages or specific performance, for neglect by the landlord to comply with fitness obligations.

---

⁶¹ *Ibid*, para 8.20.
⁶² *Ibid*, para 8.39, and Draft Bill cl 5(5), enshrining in statute the principle of *Buswell* v *Goodwin* [1971] 1 WLR 92.
⁶³ The applicable fitness standard being modelled on that of Housing Act 1985 s 604 (Report of 1996, para 8.54; draft Bill, cl 7).
⁶⁴ Report of 1996, para 8.35; draft Bill cl 5(1) and (3).
⁶⁵ For example, where temporary housing accommodation pending development of land was being provided (*ibid*, para 8.49).
⁶⁶ Report of 1996, para 8.36. The revised tenants' implied duty appears at para 10.37, and is an amalgam of the present law of waste and tenant-like user.
⁶⁷ *Ibid*, para 6.16.
⁶⁸ Report of 1996, para 6.21.

The Law Commission proposals would remove an anomaly in the law which allows tenanted houses and flats to be in repair and yet remain unfit for human habitation. They would adopt the same policy as section 8 of the Landlord and Tenant Act 1985 as regards unfitness – thus general criteria would apply to all dwellings. Rental limits would be abandoned. In the private sector, however, any increased obligations on landlords would presumably have to be paid for out of increased rents.[69] In the social housing sector, part of the bill might fall on the Treasury, which may explain the reluctance of the government to enact reform of the law along the exact lines proposed by the Law Commission.

---

[69] As pointed out by Bridge "Putting it Right" [1996] Conv 342 at 350.

Chapter 4

Repairing obligations of landlords

I – General considerations

A – Allocation of liability

Introduction

Statute apart, landlords seem to prefer not to expressly undertake the liability for repairing or maintaining demised premises. They prefer to cast the burden of repairs and maintenance on the tenant.[1] If they cannot do so directly, landlords, at least of multi-occupied premises, will ordinarily be able to keep a clear rental income by charging the tenants with the cost of works. Ordinarily, the extent of any liability of a landlord to undertake repairs or maintenance is to be discovered from the express terms of a covenant, but it may be unsafe to assume that there inquiries end. Thus, a landlord who assured sitting tenants prior to their acquiring long leases that he would execute initial roofing repairs was held to have waived any right to recover the cost under the leases granted in due course.[2]

A question has arisen as to whether a landlord could be liable, if he has expressly covenanted to give the tenant the quiet enjoyment of the premises, to carry out work to improve the design of the premises or to replace an item essential to the use of the premises by the tenant, but which has gone beyond the stage where it may be saved by tenants' repairs. The New Zealand High Court has even held that a landlord had to replace a rain pipe, lying under his land, which had reached the end of its life, because he had undertaken to give quiet enjoyment to the tenant, and so on this

---

[1] See Lewison, Ch 7; Luxton and Wilkie, Ch 10. According to "Monitoring the Code of Practice for Commercial Leases" (DETR Vol 1 para 4.2.38) some short business lessees may have succeeded in casting liability for external repairs on their landlords, although, it seems, the FRI lease still dominates the commecial sector taken as a whole (para 5.2.48).

[2] *Brikom Investments Ltd v Carr* [1979] QB 467, CA.

basis had promised to ensure a continuing supply of water to which the tenant was entitled.[3] This wide view of an express covenant for quiet enjoyment, as going beyond mere acts of physical interference with the exclusive possession of the lessee has been rejected by the House of Lords. It refused to require a council landlord to upgrade the totally inadequate sound-proofing of a flat, which had been within the relatively low building regulations at the time, to modern requirements.[4]

In the case of a lease of an independent unit to a tenant, such as a house,[5] or a shop and flat above, in a free-standing, or vertically subdivided building, the landlord may take an express obligation to keep the exterior and interior in repair from the tenant. By contrast, a landlord letting a number of interdependent units in a single building or area, such as in a shopping precinct or a block of flats with common parts such as gardens and lifts, may prefer to undertake the liability for repairs and maintenance to the fabric of the building, as well as for heating, hot water and other services for tenants, in return for the payment by each lessee of service charges. There are no doubt many variants of such clauses. In view of the propensity of service charges clauses to generate disputes, the services or items for which service charges are recoverable, not only in relation to the premises let but also in respect of those parts of the premises excluded from the demise of individual units should be comprehensively and clearly identified.[6]

Service charges – some problems

The tenant may have to pay a fair proportion of landlords' charges, or some other defined proportion of the landlords' costs. One means of apportioning landlords' costs is by fixed charges equally apportioned when the premises are let, or by reference to the net rateable value of the demised premises.[7] Although this latter method allows for the value to be fixed by a third party, it may have

---

[3] *Westpac Merchant Finance Ltd* v *Winstone Industries Ltd* [1993] 2 NZLR 247.
[4] *Southwark London Borough Council* v *Mills* [1999] 3 EGLR 35.
[5] For a term of seven years or more, owing to s 11 of the Landlord and Tenant Act 1985.
[6] Luxton and Wilkie, pp 266–267.
[7] See eg *Precedents for the Conveyancer* 5–31, cl 4.

unhappy results for lessees. Where the proportionate rateable value basis of sums due from a shop lessee was not that as appearing on the valuation list for two consecutive years, the landlord could reopen the accounts and collect the substantial additional sums, based on the rateable value in fact shown on the valuation list, from the lessee.[8] The rateable value basis may be generally unfair: the benefit of any individual unit from services or work may not correspond to the relevant part of the rateable value.[9]

Another way of fixing charges is by floor space[10] or by reference to the anticipated use by the lessee, or a by a fixed proportion of a larger total sum. The proportion of the sums due from the tenant may be based on money actually spent by the landlord, or, if clearly specified, on sums estimated to be spent by the landlord in a future accounting period. The landlord may be entitled to recover specified costs such as preparing forfeiture notices[11] and legal fees.

As pointed out by Scott J[12] in relation to a wide landlords' obligation, a service charges provision "is not ... simply ... for the benefit of the tenants. It is also a provision for the benefit of the landlord. It enables the landlord to keep its building in repair at the tenants' expense." It was also said that one advantage of this arrangement for landlords is that they keep the control over the way in which the work is to be carried out – unless the tenant has an express right to be consulted as to the execution and manner of execution of works as a condition precedent to recovery of service charges.

Reform revisited

Because of the limited circumstances in which the courts will imply repairing obligations against landlords, there may be circumstances in which neither landlord nor tenant is liable for repairs, as where

---

[8] *Universities Superannuation Scheme Ltd* v *Marks & Spencer plc* [1999] 1 EGLR 13, CA.
[9] For this and other criticisms of the rateable value method see Heighton, "Fair Means or Foul Up", *Estates Gazette* October 24 1998.
[10] See the clause suggested by Luxton and Wilkie, p 273.
[11] Special restrictions apply to a landlord's ability to enforce by forfeiture certain claims against residential tenants to service charges in Housing Act 1996, ss 81 and 82.
[12] In *Plough Investments Ltd* v *Manchester City Council* [1989] 1 EGLR 244 at 247M.

there are common parts to the demised premises or a building on which the demised premises depends. As we have noted elsewhere, the Law Commission recommended that there should be two default covenants imposed on landlords by legislation. The covenants would relate to the premises let and to common parts or property controlled by the landlord which might affect the property let. The default covenant would not apply to leases of houses and flats for a term of less than seven years, owing to the widening of the landlord's repairing obligations under the current legislation which was proposed. Oral leases would be excluded from this reform.[13]

The proposed reforms are these. The first default covenant would be a covenant by the landlord to keep the whole and each and every part of the demised premises in repair.[14] The covenant would be capable of being excluded or modified in writing by the parties, as where there was an express repairing obligation in relation to all of the relevant parts of the premises.[15] The second proposed default covenant, equally capable of being expressly excluded from the lease in whole or in part, would apply to the common parts of the building demised as well as to other parts of the premises under the landlord's control. The covenant would be appropriate where part of a building was let and part retained by the landlord, as where there are a number of different units let to different tenants and the common parts such as the common passageways and staircases have not been let to any tenants. The landlord would be required to keep each and every part of the "associated premises" in repair, to such a standard as was appropriate having regard to the age, character and prospective life of the premises and their locality.[16] However, the landlord would not be liable under the second default covenant proposed unless the tenant proved that the disrepair affected the enjoyment of the property leased to him, any common parts of the building containing his premises or any easement or licence over other property of the landlord.

---

[13] *Responsibility for State and Condition of Property* (1996), para 7.15.
[14] Including an obligation to both put and then keep the premises in repair (1996 Report, para 7.11).
[15] Report, *supra*, para 7.10.
[16] Report, para 7.27.

Comparative assessment of reform proposals

The proposals indicated to one critic that the Law Commission have "properly recognised the political expediency" of distinguishing between the commercial and the short residential sectors, for commercial landlords are "unaccustomed to statutory interference with their autonomy".[17] The Commission anticipated that in many leases, especially of business premises, the default covenant would be contracted out of by the parties.[18]

If the law is changed so as to provide a legislative fall-back covenant for all landlords save those letting on short residential tenancies, the question arises as to what difference such reforms would make. A brief look accross the Channel may be of interest (for a note about Scots law, see Chapter 1). The French Civil Code[19] imposes, in particular, two residual but important obligations on a business landlord. First, he must deliver the premises or "thing" let to the tenant. Secondly, landlords must maintain the "chose" or premises in a state suitable for the purpose for which they have been let. The end result of these obligations seems to be that the landlord is liable for initial repairs to the premises, and must then repair and maintain them in a condition suitable for the purpose for which they have been let. Thus if premises were let as an hotel, they must be suitable for occupation as such.[20] A landlord had to to repair a roof which had become seriously damaged due to prolonged lack of repair.[21] Similarly, a landlord who failed to maintain a central heating system which had broken down for some nine months committed so serious a breach of his implied obligations that the tenant was entitled to terminate the lease.[22] The implied obligations last throughout the lease – and may require French landlords to exercise constant supervision of the state of the premises.[23] As might be expected, if premises are not fit for occupation for the tenant's purpose when the lease begins, the landlord must remedy the defects – as where a lift failed and had to

---

[17] Bridge, "Putting it Right" [1996] Conv 342, pp 344–345.
[18] Report, para 7.9.
[19] Articles 1719(1) and (2) and 1720.
[20] See Dalloz, *Code Civil*, 1997/98 note to art 1720; also Civ 3e 7.1.1998. Rev Dr Immob. 1998.302.
[21] CA Paris 15.12.1994, cited in Dalloz, *supra*, p 1312 (art 1719).
[22] *Ibid*, citing CA Paris 4.5.1994.
[23] CA Montpellier 4.6.1957, D. 1957. somm.69.

be replaced.[24] The landlord can avoid liability if the extent of the work required to comply with his obligation amounts to "reconstruction" rather than "repair" – an aspect which will be familiar enough on this side of the Channel.[25]

It appears that commercial landlords in France often contract out of their basic obligations, which they are entitled to do, as the Civil Code obligations are not "public order" in nature. This practice has been attacked as an abuse of superior landlord bargaining power,[26] on the basis that many landlords are large organisations such as pension funds against small business tenants. A not untypical contracting out term in a lease may provide, for example, that the tenant takes the premises in the state in which he or she finds them, making it clear that the landlord is not liable for any repairs during the lease.[27] Another example of a contracting out clause is an undertaking by the tenant to do all except specified repairs during the lease.

Despite the freedom to contract out, the trend of cases in France appears to be hostile to contract out clauses, which are strictly interpreted. If a clause exonerates the landlord from initial repairs, it will not necessarily be treated as exempting him from carrying out major repairs during the lease.[28] Nor can a landlord necessarily avoid liability for public works by a clause casting liability for repairs on the tenant.[29] As noted elsewhere, contracting out in the residential sector is limited because the landlord cannot avoid his liability to ensure minimum norms of habitability.

French law also imposes a liability on landlords for latent defects.[30] The aim of French law is to protect tenants, who cannot discover latent defects,[31] as to which the landlord may have the best

---

[24] Cass Civ 14.10.1965, D. 1966.41. It seems that in contrast to England and Wales, a French landlord cannot demand rent if and so long as he is in breach of his civil code obligations, where applicable: Civ 3e 30.11.1994, Loy et Corpr 1995 No 162.
[25] See eg Carbonnier 1952 Rev Trim Dr Civ 149.
[26] Auque, *Baux Commerciaux* (1996), No 102.
[27] Juris Classeur, *Bail*, No 185, cf Civ 3e 10.5.1991, Rev Dr Immob. 1992.124.
[28] Cf Paris 10.1.1991, D. 1991.IR50; Civ 3e 10.5.1991, D. 1991. somm.353.note.
[29] See eg Auque, *op cit*, No 100; Civ 3e 13.7.1994, D. 1994. IR.200.
[30] Civil Code art 1721; see Auque, No 108ff.
[31] As opposed to patent defects, where it is seemingly assumed that the tenant will make his own inquiries: hence the landlord can avoid the guarantee obligation by informing the tenant of a given defect: cf note to Soc 29.3.1957, Gaz Pal 1957.2.59.

chance of being informed, until after he has taken possession. The guarantee is broken both by a state of disrepair and unfitness for use, as where land could not be used as a shop as it was formerly used to site a petrol station[32] – there seems to be no distinction between the two concepts. The guarantee obligation does not cover damage caused by fortuitous happenings such as floods. It may be contracted out of, if the clause is clear and precise,[33] save in the case of residential lettings.

If the Law Commission reforms of the general law as opposed to the short residential sector were enacted, the overall position in England might be not dissimilar to that in France. If the landlord wished to cast his statutory basic duties to repair onto the tenant, he would need to use clear language, otherwise he would remain liable for repairing work, both at the commencement of the lease and during its term, whether the cause of the disrepair was patent or latent defects. Questions would arise as to the scope and interpretation of contracting out clauses or of clauses shifting some of the landlord's liability to the tenant, and time alone would tell whether our courts would react to these clauses in a similarly hostile way to that of the French courts.

B – Interpretation of service charge clauses

The recovery of service charges is capable of being contentious. Some service charges clauses reserve the sums due as rent in advance. If the landlord is held at some subsequent date not to have been entitled to recover the whole or some part of service charges, the tenant may now seemingly bring an action in restitution for the sums overpaid, time running only from the period when the plaintiff discovered the mistake or ought, with reasonable diligence, to have discovered it.[34] This ruling may mean

---

[32] CA Versailles 14.12.1989, D. 1990. somm.259; likewise where the electrical wiring was faulty: Civ 3e 21.11.1990, Rev Dr Immob.1991.264.
[33] See eg Civ 3e 10.12.1980, Gaz Pal 1981. 1. pan 122. Some *doctrine* (eg Malaurie and Aynes, *Droit Civil, Contrats Spéciaux*, 4th edn, No 683) holds that if a landlord knows of the latent defect, he would act in bad faith if he relied on a contracting out clause, if he did not inform the tenant. It would be interesting to see, if reform arrives in England, whether the common law was prepared to go this far.
[34] *Kleinwort Benson Ltd v Lincoln City Council* [1998] 3 WLR 1095.

that if the tenant at any time overpays a service charge, he is entitled, subject to the same limitation period, to recover the excess from the landlord in a restitution action. The landlord is taken in both these cases to have been unjustly enriched at the tenant's expense. He might be able to invoke a change in his position since the moneys had been received. There might be limits to this defence. Thus the example given in the main ruling, of giving away the money, suggests that a landlord could not plead that he had spent the money on the premises in the mistaken belief that the tenant was bound to pay the charges, although it is suggested that if an assignee of the reversion did so he might make out a defence of change of position even in such a case.

The landlord may be required to supply the tenant at the end of each year of account with a copy of a certificate stating the amount of money spent on service charges and perhaps the tenant's share of the amount claimed. If the landlord's surveyor certifies that a sum is recoverable, no certificate, even if stated to be final and not capable of challenge in the courts may oust the jurisdiction of the court as to do so would conflict with public policy.[35] As a landlords' service charge clause gives him control of the way in which the work is carried out, the right to recover certified costs is subject to an implied condition precedent that these are fair and reasonable in amount,[36] so opening up a separate avenue of challenge to tenants.

A landlord or his managing agent must behave reasonably in charging for specified works for a given purpose. A landlord could not, therefore, recover all sums spent on works simply because he thought these might satisfy a stated purpose of maintaining the building as a first-class block of residential flats.[37] This landlord failed to recover for the removal of the whole brickwork skin of a building, which would amount in law to an improvement,[38] under a provision allowing him to maintain the amenities and facilities appropriate to a first-class block of flats, when he failed to recover the money under the repairs and maintenance part of the service charges clause. At the same time, the tenant must act reasonably,

---

[35] *Re Davstone Estates Ltd's Lease* [1969] 2 Ch 378.
[36] *Finchbourne v Rodrigues* [1976] 3 All ER 581, CA.
[37] *Holding & Management Ltd v Property Holding & Investment Trust plc* [1990] 1 EGLR 65, CA.
[38] In *Sutton (Hastoe) Housing Association v Williams* [1988] 1 EGLR 56 the opposite result was reached, as a matter of construction, in relation to replacement windows.

and so could not insist on the landlord doing repairs to a minimum standard, if a reasonable owner would have paid for the works in question.[39] But a landlord cannot expect to delay in the execution of necessary repairs, so increasing their cost, and then pass on the extra expense in service charges to his tenants.[40]

C – Definitions and related aspects

The extent of a landlord's express obligation to keep in repair, to repair or to maintain is governed by the definitions and surrounding circumstances of the particular lease in question.

1. *Some uncertainties* – If the landlord undertakes an express obligation to repair the structure and exterior of demised premises, there may be uncertainty as to both what amounts to the "structure and exterior" and "the demised premises", as the landlord's obligation cannot extend beyond either limit. However, repairs to the roof, main outside walls, dividing walls, inside supporting or load-bearing walls, and to the foundations of a building, would ordinarily be within the scope of "structural" repairs, or amount to repairs to the "structure and exterior" of premises. The latter term would not seemingly apply to the "many and various ways in which" premises can be "fitted out, equipped, decorated and generally made to be habitable" (or presumably, fit for any given use).[41]
2. *Windows* – If a landlord is expressly liable for repairs to the structure or exterior of premises, without further detail, the question as to his liability for windows arises. Large unopenable plate-glass windows forming most of the side of a building were held part of the structure of business premises and not within a tenant's covenant to repair landlords' fixtures.[42] By contrast, ordinary wooden-frame windows were held not to form part of the main walls and fell outside an exception to a lessee's covenant referring to the roofs and main walls.[43] Where a

---

[39] *Plough Investments Ltd* v *Manchester City Council* [1989] 1 EGLR 244 at 247M–248A.
[40] *Loria* v *Hammer* [1989] 2 EGLR 249 at 259C.
[41] *Irvine* v *Moran* [1991] 1 EGLR 261 at 262F. For a definition of "main structure", see *Hallisey* v *Petmoor Developments Ltd* [2000] EGCS 124.
[42] *Boswell* v *Crucible Steel Co* [1925] 1 KB 119, CA.
[43] *Holiday Fellowship* v *Viscount Hereford* [1959] 1 WLR 211.

landlord covenanted in long leases of flats to repair "glass windows therein", but the leases were not clear as to whether window-frames were demised to the lessees (who had to clean but not to repair the windows), the external walls of the flats were held, despite a "jumbled description" in the demise, not to have been demised to lessees: logically, therefore, the window-frames could not be and they fell within the landlord's repairing covenant.[44]

3. *Structural repairs* – Where a lease imposed on the landlord an obligation to keep the main structure and roofs of demised premises in repair, the estimated replacement of some 350 to 500 slates out of about 12,000 in total amounted to a structural repair of a substantial nature. The work could not be dismissed as a "mere trifle", since the framework of the building was to be interfered with.[45] Vaisey J[46] had indicated that the expression "structural repairs" meant "repairs of or to a structure" but this has been criticised,[47] as dividing all repairs into structural or decorative. If the expression "structural" is taken to involve "substantial" repairs, it may be that the obligation of a landlord in such a case is confined to the load-bearing elements of the relevant building, or it might imply that any serious repair is the landlord's liability.[48] If the latter approach is taken, the courts might have to make use of their "knowledge of the world", in Scarman LJ's phrase,[49] to ascertain the extent of the landlord's liability.

II – Extension of landlord liability

A – *Express guarantees*

In some cases, tenants holding new premises on long leases have been able to procure from the landlord an express term in the lease

---

[44] *Reston Ltd* v *Hudson* [1990] 2 EGLR 51; the costs incurred by the landlord were, subject to sections 19 and 20 of the Landlord and Tenant Act 1985, recoverable from the tenants.
[45] *Granada Theatres Ltd* v *Freehold Investments (Leytonstone) Ltd* [1959] Ch 592 at 605 (Jenkins LJ), CA.
[46] In the *Granada Theatres* case at first instance [1958] 1 WLR 845 at 848.
[47] Ross, 4th edn, para 8.7.
[48] *Ibid*.
[49] In *Woodward* v *Docherty* [1974] 1 WLR 966 at 969G (Rent Acts).

under which, for example, the structure and exterior are guaranteed to be in good repair and condition and free from inherent defects at the commencement of the lease. The landlord may even undertake, with or without a lessee's general covenant to keep in repair, to be bound by this sort of guarantee until the expiry of the lease. If the terms of this sort of undertaking are broken, the landlord will presumably have to restore the premises to the condition which they ought to have been in at the commencement of the lease, including curing damage caused by patent or latent "inherent" defects.[50] Thus, landlords, who built and then let a restaurant in a motel complex who covenanted "to keep the main walls and roof of the premises in good structural repair and condition throughout the term and to promptly make good all defects due to faulty materials and workmanship in the construction of the premises", were held liable to pay for the cost of the making good of severe structural damage caused by defectively built foundations.[51] A tenant who had proved breach of so fundamental an undertaking might now be able to claim that if the landlord persistently failed to put matters right, he had repudiated the lease, and the tenant could accept this repudiation by leaving the premises without liability for future rent.[52]

B – Express warranties

It has been said that a tenant who is being asked to accept a full repairing obligation for a building ought to demand from the landlord that he procure for the tenant's benefit an express warranty from the builder, for example, that the building has been put up with all due care.[53] But whereas liability for breach of a landlord's express warranty forming part of his repairing obligations is strict, requiring proof that the work is needed as a fact and that it falls within the terms of the warranty, the tenant would have to prove negligence against any builder of the

---

[50] According to Ross, 3rd edn, p 225, the expression "inherent defect" might include existing but invisible defects at the commencement of the lease resulting from eg defective design, defective workmanship or defective materials used during the construction of the building.
[51] *Smedley* v *Chumley & Hawke Ltd* [1982] 1 EGLR 47, CA.
[52] See *Hussein* v *Mehlman* [1992] 2 EGLR 87.
[53] See Ross, 4th edn, para 8.16.1.

landlord's. If the lease is granted on or after January 1 1996, it would seem that the benefit of a landlord's warranty would pass to the assignee of the lease under section 3 of the Landlord and Tenant (Covenants) Act 1995, unless the warranty was expressed to be personal to the original tenant, as it might be if the landlord wishes to limit his own risks. A builder's collateral warranty would not seem to be a "landlord covenant" within section 28(1) of the 1995 Act, so that the benefit of it would not pass automatically under the Act on an assignment of the lease, but this question has yet to be resolved.

C – Collateral contracts

Landlords may also be liable for repairs outside the terms of the lease if, prior to the lease being executed, the landlord or his agent has given the tenant a collateral contract or undertaking.[54] The tenant must prove that he would not have signed the lease without this undertaking and that it is not inconsistent with the terms of the lease – so that if the tenant is under a full repairing lease, for example, the warranty ought to relate to fitness for use, as where a landlord promised a tenant orally, the lease being silent, that the drains were in good order, which in fact they turned out not to be.[55] Proof of breach of a collateral warranty (which, if proved, becomes part of the contract of lease) entitles the tenant to rescind the lease and claim damages, with no liability for rent, within a reasonable time of taking possession. The tenant could also claim damages from the landlord, if he elected to remain in possession.[56] Where a landlord had promised certain lessees that he would rectify dampness, it was held that his proposed assignee would be bound to carry out the work.[57] The leases concerned were "old" tenancies, granted prior to January 1 1996. In respect of a collateral warranty

---

[54] See *Vincent v Bromley London Borough Council* [1994] EGCS 193.
[55] *De Lassalle v Guildford* [1901] 2 KB 215, CA.
[56] *Bunn v Harrison* (1886) 3 TLR 146, CA. As to computation of damages, cf *Berry v Newport Borough Council* [2000] 29 EG 127 (statutory warranty under right to buy).
[57] *Lotteryking Ltd v AMEC Properties Ltd* [1995] 2 EGLR 13; this result may depend on the breach having been of a continuing nature; it would not seem to render the new landlord liable for any consequential losses from the unremedied condition occurring prior to the reversion changing hands.

given by a landlord under a "new" tenancy, to which section 3 of the Landlord and Tenant Act 1995 applies, the burden of any collateral warranty would appear to be transmitted by the legislation to the new landlord, unless the undertaking were to be cast in terms which were personal to the original landlord and tenant.[58]

D – Statute

A new possibility of action by a tenant against a landlord's sub-contractor, architect or engineer has been opened up by the Contracts (Rights of Third Parties) Act 1999,[59] passed on November 11 1999 and applying to contracts as from six months from that date, and which may fundamentally affect this branch of the law.[60] By section 1(1) of the Act, a person who is not a party to the contract can enforce a term of the contract if either it expressly provides that he may enforce the contract or it purports to confer a benefit on him. It seems that if, say, a landlord and a contractor agree that the benefit of a contract for the construction of premises to be let to tenants may expressly be enforced by "all future tenants", such a term would allow any person proving he was a lessee of the premises, even though not named as such in the contract, to bring an action against a contractor who installed an inherent defect which the tenant later discovered and which he had been required by the landlord to cure. However, the 1999 Act may be contracted out of (s 1(2)). It appears that exclusion of any implication of conferral of benefits on a third party is not uncommon in commercial property contracts. Owing to the fact that institutional property investors are said to be risk averse, there may be a preference for the collateral contract means of regulating any tenants' rights.[61] If this is so, then one of the advantages of the statutory reforms, the fact that third party rights would be laid down in the primary contract,[62] is not being conferred in that sector.

---

[58] See further Fancourt, Chapter 12.
[59] See Minogue, "Deconstructing privity" *Estates Gazette* December 4 1999, 144.
[60] According to Lord Browne-Wilkinson in *Alfred McAlpine Construction Ltd v Panatown Ltd* The Times, August 15 2000, HL.
[61] Rodrigues, "Beware of Benefits by Implication" *Estates Gazette* August 5 2000, 79. See also Perks (2000) 4 L&T Rev 21.
[62] See Law Com No 242, *Privity of Contract* (1996) para 2.17.

III – Dependent and independent covenants to repair

In certain authorities concerned with leases of agricultural holdings, the tenant promised to carry out specified repairs but only provided the landlord had previously put and kept the premises let in repair.[63] If, as a matter of construction, the landlord's and tenant's covenants were independent, the latter's could be enforced even though the landlord had failed to comply with his own obligation. Equally, if the court held on the facts that the covenants of the two parties were dependent, as where it was a condition precedent to the tenant carrying out repairs that the landlord supplied the tenant with materials, the tenant was not in breach of his obligation unless the landlord first complied with his undertakings.[64]

In the more modern context of service charges obligations, similar issues have arisen. However, policy matters are here involved, notably the matter of protecting lessees against overcharging by landlords. In the absence of clear language, the Court of Appeal was not prepared to construe an obligation on lessees promptly to pay service charges as amounting to a condition precedent to the provision of services by the landlord.[65] Otherwise, a tenant might be forced to pay a service charge in prejudice of any rights to challenge individual items of expenditure. Indeed, the relevant obligation may be construed as imposing a condition precedent that unless and until the lessees are consulted, in a manner prescribed in the lease, about proposed or specified works, no works are capable of being charged for under a service charge clause. The more comprehensive the structure of any advance consultation procedures, the more likely it is that the court will hold compliance with such procedures to be a condition precedent to recovery of charges.

For example, certain landlords undertook not to accept any tenders or enter into any contracts or commission any repairs without first submitting the same with plans for the tenants' approval. External repair work to the premises was executed, with the landlords ignoring the consultation procedures, which were held to be a condition precedent to recovery. Their claim to recover

---

[63] As in *Cannock v Jones* (1849) 3 Ex 233.
[64] *Westacott v Hahn* [1918] 1 KB 495, CA.
[65] See *Yorkbook Investments Ltd v Batten* [1985] 2 EGLR 100, CA.

some £166,255 as the tenants' share of the cost failed.⁶⁶ The clear intention of the parties was that the tenants were to be consulted at every stage from estimates to contract. Likewise, a landlord who spent some £36,707 on substantial repairs to the exterior of a building was not entitled to recovery from lessees because he failed, contrary to a "peremptory" clause in the lease, to submit a copy of the specification for major or substantial repairs and estimates to the lessees before carrying out the works, which obligation was an essential part of a mechanism, triggering a right to object by the lessees and a disputes resolution procedure.⁶⁷ In both cases, the court assumed that if the parties provided for these types of procedures, there would be a sanction for non-compliance.

IV – Interpretation of landlords' express covenant

It was said in an old case that there was no difference of principle between the interpretation of a lessee's and a lessor's express repairing covenants.⁶⁸ The interpretation of individual landlords' obligations depends on all the circumstances, including the presumed expectation of the tenant in the case of a lease of a new building that it will not be out of repair nor in poor condition, so that he may use it from the date of the lease for the purposes of the tenancy.⁶⁹ The landlord may find himself having to cure the results of inherent defects for which his builder may be liable, yet not necessarily able to recover the cost from lessees, owing to his promise that the premises are, when let, in a good state of repair and condition. It is at that date, or the date the tenant of a new building takes his lease, that the landlord's obligation becomes effective.

A – *Express covenant to keep in repair*

The Court of Appeal has ruled that an unqualified covenant by a landlord to keep the whole of a building, including two office floors let to a tenant, "in complete good and substantial repair and condition" imposed a strict obligation on the landlord. He must keep

---

[66] *CIN Properties Ltd* v *Barclays Bank plc* [1986] 1 EGLR 59, CA.
[67] *Northways Flats Management Co (Camden) Ltd* v *Wimpey Pension Trustees Ltd* [1992] 2 EGLR 42, CA.
[68] *Torrens* v *Walker* [1906] 2 Ch 166 at 174.
[69] *Credit Suisse* v *Beegas Nominees Ltd* [1994] 1 EGLR 76.

the premises in all times in repair. The landlord was in breach of covenant immediately a defect appeared. If, however, the defect occurred within the demised premises, it was a condition precedent to liability to remedy the defect that the landlord is given notice of the lack of repair by the tenant (or a third party).[70] On the facts, the landlord was liable in damages to the tenant, despite his lack of notice, the damages being computed as from the date when the defects first appeared, some 18 months before the landlord finally remedied them.[71] No concluded view was expressed by Nourse LJ[72] "as to the case where a defect is caused by an occurrence wholly outside the landlord's control". The example cited was of a roof being damaged by a branch from a tree standing on neighbouring property so that rain water entered the tenant's premises by that means. By contrast, where an external downpipe on the landlord's premises was blocked by a dead pigeon which the landlord failed to clear away, he was held liable to his lessee for damage caused.[73]

Nourse LJ did not think that many landlords entered into express obligations merely to repair.[74] However, should a question arise as to whether a landlord who covenants simply to repair is liable first to put premises into repair and then to keep them in a good state of repair, then by analogy with two Canadian authorities, landlords could claim that an obligation to repair is limited to repairing the premises, during the lease, having regard to their state and condition when let. On this view, they would not be required to undertake the further liability, in the case of premises which are out of repair at the start of the lease, to remedy initial dilapidations.[75]

A landlord's obligation to keep in repair extends to subordinate renewal. An obligation to maintain on its own or in conjunction with an obligation to repair connotes that the landlord will keep the premises in their existing form. He may have to use different

---

[70] *British Telecom plc* v *Sun Life Assurance Society plc* [1995] 2 EGLR 44.
[71] See also *Passley* v *London Borough of Wandsworth* (1996) 30 HLR 165, CA. In *Bavage* v *Southwark LBC* [1998] CLY 3623, a landlord was accordingly held liable for damage caused to a tenant from a vandalised sewage stack retained by the landlord. This result emphasises the need for regular inspections to guard against the results of sudden damage.
[72] In the *British Telecom* case, *supra*, at 46L and 629G.
[73] *Bishop* v *Consolidated London Properties Ltd* (1933) 102 LJKB 257.
[74] In the *British Telecom* case, *supra*, at 46M.
[75] *Manchester* v *Dixie Cup Co* (1952) 1 DLR 19 at 31 (Roache); *Weinberg Estate* v *Intext Linen Supply Ltd* (1992) 24 RPR 75 at 83.

materials to those used to construct the damaged item. He does not have to alter the basic design of the premises. An obligation to repair includes an obligation not to destroy.[76] On the other hand, if the landlord has undertaken to keep a heating system in good repair and condition, it is no answer for him to install a cheaper system if the original system fails and is capable of being replaced with a similar system, since like must be replaced with like or the equivalent where possible, even if some improvements in design are involved.[77]

B – Rebuilding

A covenant to keep in repair which does not in terms require the landlord to rebuild the whole premises, does not require the landlord to remedy the defects in question if the premsies have reached a condition where they cannot sensibly be kept in repair. Three questions may be asked in this connection:[78]

(1) Whether the alterations go to the whole or substantially the whole of the structure or to a subsidiary part.
(2) Whether any alterations (or other work) produce a building of a wholly different character to the building let.
(3) As to the cost of the works as related to the previous value of the building and their effect on its value and lifespan.

The question of whether work which a landlord is seeking to carry out and then charge to lessees is rebuilding or renewal or subordinate renewal is largely one of fact and degree, as it would be with a lessee's covenant. Where the steel frame of a building which was 50 years old at the date of the grant of the leases was rusty, the lessor produced a scheme to cure the damage, at a cost of between £383,553 and £507,990, but a declaration that such work fell within the lessor's covenant to repair the exterior was refused.[79]

---

[76] *Devonshire Reid Properties Ltd* v *Trenaman* [1997] 1 EGLR 45 at 46C.
[77] See *Creska Ltd* v *Hammersmith and Fulham London Borough Council* [1998] 3 EGLR 35 (tenants' liability to replace under-floor heating system with latest equivalent type of system and not by night-storage heaters).
[78] *McDougall* v *Easington District Council* [1989] 1 EGLR 93 at 96A, CA.
[79] *Plough Investments Ltd* v *Manchester City Council* [1989] 1 EGLR 244. The landlords were entitled to the costs of a survey directed at remedying disrepair but not to the costs of a complete structural survey.

The steel frame was already in a damaged state when the leases of this mature building were granted. By contrast, the High Court held that landlords of a commercial building occupied by a number of office lessees who were under an obligation to keep in good repair and maintain the window frames were entitled to replace the existing, and damaged, single-frame steel window-frames with aluminium, double-glazed windows.[80] The change in materials and quality was not so significant as to transform the work from repairing to renewal work. Equally, where the cost of replacing wooden-frame windows to a block of flats with maintenance-free metal frame windows was double that of putting the existing windows into proper repair, it was held that it would be an improvement to install these new windows, which would be different in kind to the existing ones. The cost of so doing would be irrecoverable from the lessees.[81]

C – Improvements

A landlord's covenant to keep in repair does not, in the absence of clear language, enable or require the landlord to improve the design or condition of the premises beyond their state at the date of the grant of the lease. In one case,[82] the landlord had covenanted to keep the external part of the premises in good tenantable repair and condition, but was held not liable to insert a damp-proof course into the demised cellar, which was 100 years old when let and never had such a feature. He was only obliged to make good by repointing the external brickwork. Likewise, a council landlord was not bound to insert a damp-proof course into an old house which had never had such a feature.[83] Owing to the principle of replacing like with like under a repairing obligation, a landlord of a flat was held obliged to replace a failed damp-proof course with a new one of the most modern and practicable type, despite the

---

[80] *Minja Properties Ltd* v *Cussins Property Group plc* [1998] 2 EGLR 52, CA.
[81] *Mullaney* v *Maybourne Grange (Croydon) Management Co Ltd* [1986] 1 EGLR 70.
[82] *Pembery* v *Lamdin* [1940] 2 All ER 434, CA; likewise, a landlord was not required to insert a new rose and cover at the top of a drain where these were not part of the original design of the premises: *Masterton Licensing Trust* v *Finco* [1957] 76 NZLR 1137, CA (NZ).
[83] *Wainwright* v *Leeds City Council* [1984] 1 EGLR 67, CA.

slight small-scale improvement involved in the work.[84] Thus, if the landlord is being asked or is himself trying to alter the original construction of the demised premises, the work is likely to be classified as an improvement. The difficulty of drawing the line is shown by the fact that whereas the High Court concluded that a landlord could recover from lessees the cost of installing a new storm-proof roof to prevent a roof collapse after the 1987 storms, the tenants having agreed to pay for works of "renewal or replacement",[85] where landlords installed a new, multi-faceted roof with new design features such as a waterproof membrane, the Ontario Court of Justice concluded that as a matter of degree the work went outside a covenant to repair.[86]

D – Disrepair

To trigger liability under a covenant to keep in repair or to repair, there must be physical damage to the item in question before the landlord can be obliged to carry out any repairs.[87] If, in the process of curing a state of disrepair, an inherent fault was cured, the landlord could not escape liability provided the work overall is classified as a repair.

E – Clean-up of contaminated land

Under Part IIA of the Environmental Protection Act 1990, there is a potential liability on an "appropriate person" in consequence of the presence of contaminating substances found in or under land which has been designated as being contaminated by a local authority.[88] A landlord, as an "owner or occupier" is an "appropriate person"

---

[84] *Elmcroft Developments Ltd* v *Tankersley-Sawyer* [1984] 1 EGLR 47, CA; also *Alexander* v *Lambeth London Borough Council* [2000] 2 CL 386 (underpinning part of premises constituted a repair).
[85] *New England Properties Ltd* v *Portsmouth New Shops Ltd* [1993] 1 EGLR 84.
[86] *Shenkman Corpn* v *OAC Holdings Ltd* (1997) 8 RPR 3d 118.
[87] *Post Office* v *Aquarius Properties Ltd* [1987] 1 EGLR 40, CA ("a remarkable case", where intellect perhaps took precedence over any other process, *per* Harman J in *Minja Properties Ltd* v *Cussins Property Group plc* [1998] 2 EGLR 52 at 55F).
[88] For details of the contamination designation and remediation procedure see *Encyclopedia of Environmental Law and Practice*, Vol 2 para 2052ff; also Tromans, *Commercial Leases*, p 347ff.

within section 78F(4) and (5) of the 1990 Act, where the person causing any "historical" contamination cannot, after reasonable inquiry, be found. The landlord may then be served with a remediation notice by the local authority concerned (s 78E). He may have to pay for the reasonable costs of cleaning up the land. A question arises as to whether such costs fall within the scope of a landlord's covenant to keep in repair, and so *prima facie* within the terms of any service charges clauses. If the lease does not contain an exclusion or limitation clause in relation to historical contamination, it is difficult to see how a tenant could, as a matter of principle, be liable to pay for the results of contamination as repairing works. A covenant to keep in repair is distinct from a covenant to cleanse.[89] While remedying physical injury to any buildings caused wholly or partly by contamination might well fall within a landlord's covenant to keep in repair, such a covenant does not seem ordinarily extend to any underlying land.[90] The cleaning up of historical contamination would in any event seem to be an improvement to the demised property, if it was let in a contaminated state, and so ordinarily outside any covenant to keep in repair. Some recently drafted leases may, however, expressly oblige tenants to pay for the cost of clean-up operations whether or not the cost of doing so is a repair.

V – Requirement of notice of want of repair

A – *General principles*

Where a landlord is under a covenant, express, implied or imposed by statute, he must have notice of the fact of disrepair if the defect concerned occurred within the demised premises.[91] Unless and until notice is given, the landlord is not liable either to execute remedial work or in damages. If premises have deteriorated to such an extent that they cannot be repaired, the notice rule may allow a landlord without prior notice to avoid liability.[92] Certainly, the rule limits the quantum of tenants' recoverable damages, liability running only from the time notice is given to the landlord.

Notice is actual information of the fact of disrepair conveyed to

---

[89] *Starrokate Ltd* v *Burry* [1983] 1 EGLR 56, CA.
[90] Ross, 4th edn, para 7.31.1 and 2.
[91] *British Telecom plc* v *Sun Life Assurance Society plc* [1995] 2 EGLR 44, CA; *Passley* v *London Borough of Wandsworth* (1996) 30 HLR 165, CA.
[92] Law Com Consultation Paper (1992), para 3.28.

the landlord, rather than a formal notice to landlords – which latter would no doubt suffice. The notice principle is open to criticism.[93] It is difficult to reconcile with principle, since it reads limitations into an unqualified obligation of the landlord to keep in repair. The notice exception is founded on the principles that:

(1) the tenant has exclusive possession of the demised premises;
(2) he is said to be best placed to discover disrepair; and
(3) the House of Lords baulked at forcing landlords to carry out continuous inspections for what may be small-scale defects.[94]

Yet short residential tenants, who are often affected by this rule, are scarcely likely to be any more able than landlords to discover latent defects, faults in lifts and so on, whereas the notice requirement may have made sense in relation to agricultural tenants, who might have been expected to be handymen. The notice principle as it applies to the landlord's statutory repairing covenant is difficult to reconcile with the fact that the landlord has a statute-implied right of entry and inspection. However, the notice principle is not unique to England.[95]

The rule has been limited as it requires notice only of the fact of disrepair, not as to any details of remedial work. The landlord need not be informed of the precise nature or extent of the defect.[96] While at one time, only notice from the tenant sufficed, notice was more recently held sufficiently given in a local government officer's assessment of the cleanliness of the premises for the purposes of environmental health,[97] by a surveyor pursuant to a tenant's claim to buy the freehold,[98] and in a local authority statutory repair notice served on the landlord.[99]

B – Scope of rule

The notice rule extends to defects in the premises which existed prior to the granting of the lease which manifest themselves only

---

[93] See [1997] Conv 59 (PF Smith).
[94] *McCarrick v Liverpool Corporation* [1947] AC 219, at 236 (Lord Porter); 239 (Lord Simmons) and 231 (Lord Uthwatt).
[95] See *Chatfield v Elmstone Resthouse Ltd* [1975] 2 NZLR 269.
[96] *Al Hassani v Merrigan* [1988] 1 EGLR 93, CA.
[97] As in *Dinefwr Borough Council v Jones* [1987] 2 EGLR 58, CA.
[98] *Hall v Howard* [1988] 2 EGLR 75, CA.
[99] *McGreal v Wake* [1984] 1 EGLR 42, CA.

during the lease.¹⁰⁰ It also applies to defects, latent or patent, which arise and manifest themselves during the currency of the lease. If a latent defect, of which the landlord has no notice, becomes patent, the landlord is not liable in damages to the tenant. Where the ceiling of the bedroom in a subdivided building suddenly fell in owing to unseen damage caused by occupiers of the floor above, the landlord was held not liable to the tenant for injuries sustained as no notice had been given to him of the existence of this latent defect – even though the tenant could not possibly have notified the landlord of the defect prior to the collapse of the ceiling.¹⁰¹

C – Position once notice is given

Once the landlord has notice of a want of repair, then he has a reasonable time within which to execute the works in compliance with his obligations. He must ascertain the nature and extent of the works required and press on with the work. It may be that he cannot discharge his liability, after notice, merely by instructing a contractor to do the work.¹⁰² The exact amount of time allowed for these purposes will depend on the facts and circumstances of the case.¹⁰³ In the case even of emergency works, a tenant may not necessarily, especially if he is temporily out of occupation, be able to insist on immediate or near-immediate execution of works by the landlord.¹⁰⁴

VI – Landlord's right of entry to carry out repairs

A – General rules

Statutory rights of entry¹⁰⁵ and other exceptional cases apart, if there is no express reservation in the lease entitling the landlord to enter the premises and execute repairs or other prescribed work, the

---

[100] *Uniproducts (Manchester) Ltd* v *Rose Furnishings Ltd* [1956] 1 WLR 45.
[101] *O'Brien* v *Robinson* [1973] AC 912.
[102] *Chatfield* v *Elmstone Resthouse Ltd* [1975] 2 NZLR 269 at 274.
[103] *Griffin* v *Pillett* [1926] 1 KB 17. In eg *McGreal* v *Wake* [1984] 1 EGLR 42, a period of eight weeks from notice to the landlord was held a reasonable time.
[104] *Morris* v *Liverpool City Council* [1988] 1 EGLR 47, CA.
[105] Eg under Landlord and Tenant Act 1985, s 11(6); Rent Act 1977, s 148 (protected tenancies); Housing Act 1988, s 16 (assured tenancies).

Repairing obligations of landlords

landlord cannot enter the demised premises to carry out landlords' repairs without the licence of the tenant.[106] A landlord without any express right to enter and repair in a lease was held not entitled to a mandatory injunction to enter and repair, so as to overcome the lessee's resistance, even though the tenant was said to be in breach of his covenants to repair.[107]

B – Entry and repair

Some leases contain an express term entitling the landlord to enter, inspect, and execute tenants' repairs and also to charge the tenant with their cost. In such cases, the costs will be treated as a contract debt and the Leasehold Property (Repairs) Act 1938 (which applies only to damages claims) is neatly circumvented.[108] At the same time, if the tenant refuses access to the premises for the purpose of allowing the landlord to carry out repairs which the tenant has failed to execute, it cannot be assumed that the landlord would *ipso facto* be entitled to an injunction to force the tenant to submit to his entry even if the work falls within the tenant's covenant to repair. The landlord must prove diminution in the value of his reversion, or that damages are not an adequate remedy. A landlord was thus refused an injunction, in discretion, to support his express right of entry where the tenant was prepared to execute repairs to three floors but not the ground floor of certain premises. There, his mainframe computer was housed. The court held, in the very special circumstances, that it would be oppressive and disproportionate to the relatively small injury to the landlords' right to have repairs carried out to require the tenant to go to the expense and disruption of moving its computer equipment out, when the lease was soon to expire.[109]

C – Implied rights of entry

If the landlord fails to reserve any right to enter, inspect and execute landlords' repairs, and if he has no statutory right to enter

---

[106] *Stocker v Planet Building Society* (1879) 27 WR 877, CA.
[107] *Regional Properties Ltd v City of London Real Property Ltd* [1981] 1 EGLR 33.
[108] *Jervis v Harris* [1996] 1 EGLR 78.
[109] *Hammersmith and Fulham LBC v Creska (No 2)* [2000] L&TR 288. It appears that the tenants accepted that at the end of their lease, they would complete the work.

and repair,[110] he has an implied licence from the tenant to enable him to enter, reasonable times, to execute the repairs in question. The landlord, before entering, must inform the tenant of the nature of the proposed work, but furnishing the tenant with a builders' estimate was held to suffice in one case and a contention that the landlord must supply a full works specification was rejected.[111] The licence allows the landlord to remain on the premises for a reasonable time to execute the repairs. Entry under any licence is limited to that which is strictly necessary in order to do the work. The terms of the licence are a question of fact.[112] There is no implied obligation by the tenant to give the landlord exclusive occupation unless this is unavoidably necessary to enable the repairs to be done, as where the disrepair is so serious that occupation of the whole premises is the only reasonable way in which the landlord can carry out the work.[113] Where a tenant prevented the landlord from carrying out roofing repairs, so putting it out of his power to execute the work, the landlords apparently ceased to be in breach of covenant from that moment on.[114]

---

[110] In the case of a weekly tenancy a landlords' right of entry and repair is, exceptionally, implied: *Mint v Good* [1951] 1 KB 517, CA.
[111] *Granada Theatres Ltd v Freehold Investments (Leytonstone) Ltd* [1959] Ch 592 at 611.
[112] *McDougall v Easington District Council* [1989] 1 EGLR 93, where a licence was implied to permit the landlord to execute consequential decorations following major refurbishment works on terms of payment of £50 to each tenant towards redecoration.
[113] *Saner v Bilton* (1878) 7 ChD 815.
[114] *Granada Theatres Ltd v Freehold Investments (Leytonstone) Ltd, supra* at 612 (Romer LJ) and 615 (Ormerod LJ).

Chapter 5

Express repairing obligations of tenant

I – Introductory and general aspects

A – Allocation of responsibility for maintenance

The responsibility for repairing and maintaining demised premises is, at common law, allocated in accordance with the bargain made between the landlord and tenant in a lease. Except in special cases, such as short residential leases, Parliament does not intervene. Traditionally, commercial landlords seek to impose liability for underaking repairs and maintenance on tenants.[1] This may be done directly, by means of a covenant to repair from the tenant, applying to the whole of the premises demised. In the case of a building such as a shopping centre or office block, however, it is often the case that landlords undertake the responsibility for external and structural repairs and maintenance, but in return for an undertaking by each tenant to pay a portion of the cost as service charges. The latter way of dealing with repairs and maintenance has already been discussed. In this Chapter we examine various aspects arising where the tenant has expressly undertaken liability to carry out or to pay for for repairs,[2] either for the structure and exterior as well as for the interior of his premises, or only, as in the case of multi-occupied buildings, in relation to the interior of the unit demised to him.

B – Some preliminary aspects

A covenant to repair is ordinarily co-extensive with the physical extent of the premises demised to the tenant, at least where a self-

---

[1] For a survey of practice in the 1990s, see "Monitoring the Code of Practice for Commercial Leases" DETR (2000), *passim*.
[2] However, the Law Society Business Lease (Whole Building) cl 5.2, refers only to an obligation on the tenant to "maintain" the premises. As to "maintain" see *Graham v Markets Hotel Pty Ltd* (1943) 67 CLR 567.

contained unit has been let. Hence a lessee's repairing obligation will not apply to some part of a building, such as the roof, if the court finds that, as a matter of fact and degree, the roof has been excluded from the demised premises.[3] Where a demise was limited to the "underside of" a floor, it was held not to include the columns and thus girders.[4] The court examines the premises clause in a lease in the same way as it would any other clause, and so has regard both to the covenant to repair and against alterations.

Once a tenant has undertaken an express obligation to repair, the obligation operates as from the date of the lease. The tenant will have to carry out repairs even if the cause of the work is a fire or other extraordinary event such as a flood.[5] The tenant may therefore wish to procure the insertion in the lease of a suspension of rent clause to relieve him from liability to pay rent, but care must be taken with the language of the clause if it is to apply to a liability to pay service charges as opposed to rent.[6]

A covenant to repair ceases to apply with the termination of the lease, whether by effluxion of time, surrender express or implied or the operation of forfeiture by the landlord. In the case of forfeiture, there is a difficulty caused by the fact that a lease is notionally forfeited once a writ claiming forfeiture is served, whereas it is in fact only possible for the landlord to repossess the property where he proceeds by an action for possession once the court has ordered the tenant to hand back the possession to the premises to the landlord. However, it has been recently assumed that liability under any lessees' covenants continues until such time as the lease has been finally put to an end by the court ordering possession, so overcoming this problem.[7]

Sometimes, in addition to an obligation to keep in repair during the term of a lease, the lessee is further required to deliver up the premises in good repair and condition at the end of the lease, however determined.[8] Although under a covenant to keep in repair

---

[3] An unqualified demise of a building was held to include the roof: *Straudley Investments Ltd* v *Barpress Ltd* [1987] 1 EGLR 69.
[4] *London Underground Ltd* v *Shell International Petroleum Co Ltd* [1998] EGCS 97.
[5] *Redmond* v *Dainton* [1920] 2 KB 256.
[6] *P&O Property Holdings Ltd* v *International Computers Ltd* [1999] 2 EGLR 17.
[7] *Maryland Estates Ltd* v *Bar Joseph* [1998] 2 EGLR 47, CA.
[8] See eg Luxton and Wilkie, p 147; also *Precedents for the Conveyancer*, 5–31 cl 5(6).

a tenant is liable at all times to keep the premises in repair and is in default as soon as they are not in that condition,[9] whatever the cause, be it fortuitous damage or tenant fault, or an inherent defect causing physical damage, these sweeping up clauses are still being suggested, perhaps because until recently it was not clear that an obligation to keep in repair[10] was strictly construed.

Complications have arisen where a tenant holds over after expiry of his common law term. Where a fixed term tenant holds over as a periodic implied yearly tenant, his covenant to repair continues at common law into the periodic tenancy.[11] By contrast, where a business tenant holds over under statutory continuation of his tenancy under section 24 of the Landlord and Tenant Act 1954, the House of Lords refused to extend the obligation of the contractual tenancy to pay rent into the continuation tenancy. It is presumed that this ruling would apply equally to the continuation tenant's repairing obligations.[12]

C – Brief contrast between "old" and "new" tenancies

Although the general doctrines applying to the enforcement of leasehold covenants against an original tenant who has assigned the lease to an assignee are outside the scope of this work,[13] it should be noted that if a tenant holding an "old tenancy"[14] assigns his term to an assignee, he and any guarantor of his, in the absence of release, remains at risk of paying damages to the landlord for the time being if the assignee defaults in relation to his repairing covenant during the remainder of the term of the lease. By contrast, in principle, a tenant under a lease granted on or after January 1

---

[9] *British Telecom plc* v *Sun Alliance Society plc* [1995] 2 EGLR 44, CA.
[10] As opposed to a mere covenant to repair, which triggers liability in damages only after a reasonable period to execute remedial works has expired.
[11] *Ecclesiastical Commissioners* v *Merrall* (1869) LR 4 Ex 162.
[12] *City of London Corporation* v *Fell* [1994] 2 EGLR 131. It is otherwise if there are express and clear words of extension of the common law covenants into the continuation tenancy.
[13] See eg Davey, *Landlord and Tenant Law* (1999) Ch 7; Evans and Smith, *Law of Landlord and Tenant* (1997) Ch 5; Fancourt, *Enforcement of Landlord and Tenant Covenants* (1997).
[14] ie a lease granted prior to the commencement of the Landlord and Tenant (Covenants) Act 1995.

1996 is released by statute[15] from future performance of or liability under a covenant to repair, the performance of which is the responsibility of the current assignee alone.[16] A question has been raised,[17] given that the original tenant remains liable to the landlord for damages for disrepair down to the date of the assignment, the operative date of the statutory release, as to the position where a landlord brings an action against the assignee for a breach of a repairing covenant. If the landlord recovers damages from the assignee, there is nothing in the Act to prevent his recovering damages from the erstwhile tenant without giving credit for sums paid by the assignee arising out of the continuing breach. The silence of the Act is as curious as the fact that it has to be assumed that where an assignee reassigns, he is entitled as much to the benefit of a statutory release as was the original tenant – section 5 of the 1995 Act does not in terms so provide as it refers only to "the tenant".

D – Limits to repairing covenants

There are some limits inherent in the juridical nature of a repairing covenant. In particular, a covenant to repair impliedly excludes any obligation to improve the condition of the premises, save to that extent necessarily required in the act of making good physical damage.[18] This particular line is difficult to draw. It has sometimes arisen where landlords of a block of flats or office units seek to carry out work of replacement of windows or other items at a cost objected to by the tenants. Since on first principles a "repair" is going to have to cure a pre-existing condition of physical damage to the premises,[19] the premises will be improved to the extent that new materials of suitable modern quality will have been added to them. That type of work is no less a repair. If the tenant is asked to replace like not with like but with some other form of materials,

---

[15] Subject to the effect of any authorised guarantee agreement taken from the lessee where this is allowed, as to which see *Wallis Fashion Group Ltd v CGU Life Assurance Ltd* [2000] 27 EG 145.
[16] Landlord and Tenant Act 1995, s 5, which cannot be contracted out of in the lease (s 25).
[17] By Fancourt, para 14.09.
[18] *Quick v Taff-Ely Borough Council* [1985] 2 EGLR 50, CA.
[19] *Calthorpe v McOscar* [1924] 1 KB 716, CA.

which may be of better quality, as by replacing wooden frame windows with double-glazed windows, this work was characterised as an improvement and so outside a covenant to repair.[20] By contrast, it was held that the replacement of corroded and warped window frames with new frames of a different design, which were rust-proofed, amounted to a repair.[21] The different results in these two cases show the difficulty of drawing the line: in the second case the windows were badly out of repair. In one, the works proposed were a sensible method of remedying the disrepair arising, whereas in the first of the two cases, the new windows would have cost almost double the price of the current ones, which could seemingly be repaired at a lower cost. It is open to the parties to modify these principles by clear express language. Thus, a right to carry out such additional works as the landlord considered necessary was held to extend to his replacing wooden window frames with new double-glazed UPVC windows.[22]

A further question arises where the tenant claims that the work which the landlord is asking him to carry out will benefit the landlord for some time after the expiry of the lease. The issue is as to whether such work would be outside the scope of a tenants' repairing covenant for that reason. It seems that if an item requires repairing at a cost which is reasonable in all the circumstances, the fact that it might have a much shorter commercial life to the tenant cannot limit his liability to work to secure the item only for that period. Thus it was held that where a badly-built tower (the defect not being known to either party when the lease was granted) needed repairs at a cost of some £500,000, the fact that it might be secured for the tenant's purposes for a 15-year period at a much lower cost did not avail the tenant, who was assumed to have to pay in full for the necessary works.[23] As a result of this reasoning a tenant would seem to be liable to keep in repair a building which had now become redundant, such as useless dock buildings and cranes or, for that matter, "Mr George's shooting gallery and fencing school ... as described in 'Bleak House'".[24]

---

[20] As in *Mullaney* v *Maybourne Grange (Croydon) Management Co Ltd* [1986] 1 EGLR 70.
[21] *Minja Properties Ltd* v *Cussins Property Group plc* [1998] 2 EGLR 52, CA.
[22] *Sutton (Hastoe) Housing Association* v *Williams* [1988] 1 EGLR 56, CA.
[23] *Ladbrooke Hotels Ltd* v *Sandhu* [1995] 2 EGLR 92 (a rent review dispute).
[24] Per Warner J in *Ladbrooke Hotels Ltd* v *Sandhu, supra* at 95E.

II – Construction of repairing obligations

A – *General considerations*

The general approach of the courts to the construction of any repairing obligation of a tenant is to examine the particular language used in the covenant with which the court is confronted. Covenants suggested by the authors of precedents often speak of an obligation on the tenant to "keep" the premises "in good repair" or in "good and substantial repair and condition".[25] The latter form of covenant involves, according to Windeyer J, a qualitative obligation (good) and a quantitative obligation (substantial). It conveys a requirement that repairs must be undertaken to such a degree that the building is put in the condition it would have been in if good and substantial repair had been undertaken during the term of the lease.[26]

Sometimes the tenant's obligation extends to cleansing the premises, and it is often the case that the tenant, where he is holding a self-contained unit rather than a unit in a multi-occupied building, is obliged to carry out painting and decorating at regular intervals. If the tenant holds a long lease, rather than a term for a short or medium period, it may be that he would be not only subject to a covenant to keep in repair and good condition but in addition required to rebuild or renew where necessary. If indeed the tenant has undertaken onerous obligations of this kind, the courts appear willing to construe them literally, even though "in the ordinary way a covenant in a lease to rebuild the entire premises would be unusual".[27]

It may be that "draftsmen frequently use many words either because it is traditional to do so or out of a sense of caution so that nothing which could conceivably fall within the general covenant which they have in mind should be left out".[28] However, each separate word or expression in a covenant is given its full meaning,

---

[25] See eg Lewison, pp 198–199; Luxton and Wilkie, p 242; *Precedents for the Conveyancer*, 5–5 cl 2 (interior only) and 5–31 cl 5 (exterior and interior).

[26] Windeyer J's view was adopted in *Alcatel Australia Ltd v Scarcella* (1998) 44 NSWLR 349 at 354–356 (where relevant English authorities were followed).

[27] *Norwich Union Life Assurance Society v British Railways Board* [1987] 2 EGLR 137 at 138F.

[28] Per Hoffmann J in *Norwich Union Life Assurance Society v British Railways Board, supra*.

and proper effect must also be given to the context of the covenant.[29] For example, where a landlord's obligation contained eight verbs including an obligation to maintain repair amend renew and to cleanse, the words "amend" and "renew" were held to be capable of going beyond a "repair", requiring extensive work to be executed.[30] Similarly, where a tenant's service charge clause was linked to an obligation by the landlord to "renew or replace" which followed an obligation to keep in repair, the tenant was under an obligation to pay for part of the cost of the replacement of a roof which had been severely damaged in a storm with a new, different, roof of an improved design which would better withstand wind lift.[31] The High Court reached this conclusion even on the hypothesis that it might have been "improbable in the extreme" that the tenants would have agreed to pay for such work, because the words "renew or replace" were clear and not capable of being misread by the tenant's advisers. Certainly, a tenant cannot escape from his repairing obligation on the ground of mere cost. In one case,[32] a tenant undertook to keep in repair a unit including, in a list of items, the roof. The whole of the roof had failed and had to be replaced at a cost of upwards of £84,000. The High Court held that the cost of this work fell plainly within the scope of the repairing obligations of the lessee. The roof was not the whole of the building: when the work was done the landlords would have the same building with a new roof. It made no difference in this that the value of the unit as repaired was between £140,000 and £150,000. Although the cost of work is a relevant factor in some cases, the literal interpretation of the covenant in this case, combined with the fact that the work was not new and different in character as compared to the premises when let was decisive against that lessee.

The language of the particular words used in a given covenant is examined first, to ascertain its proper meaning.[33] The courts do not

---

[29] Per Beldam LJ in *Fincar SRL* v *109/113 Mount Street Development Co Ltd* [1999] L&TR 161 at 170 (a long lease): also *Welsh* v *Greenwich London Borough Council* [2000] 49 EG 118 (short tenancy).
[30] *Credit Suisse* v *Beegas Nominees Ltd* [1994] 1 EGLR 76.
[31] *New England Properties Ltd* v *Portsmouth New Shops Ltd* [1993] 1 EGLR 84.
[32] *Elite Investments Ltd* v *TI Bainbridge Silencers Ltd* [1986] 2 EGLR 43.
[33] Literal meanings of expressions as to the timing of work are not adopted: thus a covenant to put premises in repair "forthwith" was taken to allow a reasonable time for the tenant after entry to comply: *Doe d Pittman* v *Sutton* (1841) 9 C&P 706.

look at repairing obligations in isolation from the rest of the lease. It now seems that they also take into account commercial common sense, so allowing a departure from the natural meaning of a covenant to arrive at a more sensible conclusion, although the High Court noted the risk of doing so: the parties may have good reason for agreeing to a surprising result.[34]

The court will have regard to the context of the repairing covenant – as it does with any covenant. Hence, the length of the lease and the type of premises are relevant factors,[35] as well as the state of the building at the date of the lease,[36] and the contemplation of the parties when the lease was granted. The court will also have regard to the nature and extent of the defect and the cost and nature of the work proposed to put it right. Where appropriate, it takes into consideration the likelihood of recurrence if one remedy rather than another is proposed.[37] The weight to be given to each of these factors depends on the facts and circumstances of each case. It is for this reason that decisions on seemingly identical words used in an earlier lease are only of limited value.

B – Meaning of repair

1. The concept of "repair" requires consideration of what is meant in law by "repair", then of proof of disrepair, the fact that repair does not include any obligation to improve, and then some mention of the standard of repair required of a tenant under a repairing obligation.
2. The notion of a "repair" is a limited one. There must be part of the premises which is physically damaged and in need of replacement with materials of at least the same quality.[38] In other words a "repair" is the replacement of something already there, which has become dilapidated or worn out by its modern equivalent.[39] The extent of any work required by an obligation

---

[34] *Holding & Barnes plc* v *Hill House Hammond Ltd* [2000] L&TR 428; as to commercial common sense see esp. *Mannai Insurance Co Ltd* v *Eagle Star Life Assurance Co Ltd* [1997] AC 749; see further Morgan (1999) 3 L&TR 88.
[35] *Lurcott* v *Wakely and Wheeler* [1911] 1 KB 905 at 915, CA.
[36] As in *Eyre* v *McCracken* (2000) 80 P&CR 220, CA.
[37] *Holding & Management Ltd* v *Property Holding & Investment Trust plc* [1990] 1 EGLR 65 at 68H, CA.
[38] *Calthorpe* v *McOscar* [1924] 1 KB 716, CA.
[39] *Morcom* v *Campbell-Johnson* [1956] 1 QB 106 at 115 (Denning LJ) (when contrasting repairs and improvements).

to repair depends on any qualifying words used in the covenant, such as "good and substantial" – as already noted.
3. If the work involves replacing like with like, it remains classified as a repair within the tenant's covenant, even if the tenant claims to have a cheaper and equally efficacious method of solving the defects, since he is bound to repair the existing items if this is still practicable. Thus, a tenant of office premises was held liable to make extensive repairs to an underfloor heating system, including replacing a proportion of the underground cables which operated it. The existing system could be practicably maintained in good working condition – albeit with extensive repairs. The tenant was not therefore entitled to install cheaper night-storage heaters.[40] If it is impracticable or impossible to repair the existing items by equivalent replacement, the tenant would be entitled to put forward a suitable alternative scheme of work. In one case, it was accordingly held that the replacement of a damaged wooden door with a self-sealing aluminium door, was a mode of replacement which a sensible person would have adopted.[41] In some cases, it may be appropriate for the tenant to pay for work which has the effect of making further continual repairs unnecessary, provided the additional initial cost of acting in this anticipatory way is not unreasonable.[42]
4. Repair includes subordinate renewal but not the complete rebuilding or renewal of the demised premises. This is because repair is restoration by renewal of subsidiary parts of a whole. Renewal, as distinguished from repair, "is reconstruction of the entirety, meaning by the entirety not necessarily the whole but substantially the whole subject-matter under discussion."[43] The line between repair and renewal in the sense here set out is

---

[40] *Creska Ltd* v *Hammersmith and Fulham London Borough Council* [1998] 3 EGLR 35, CA.
[41] *Stent* v *Monmouth District Council* [1987] 1 EGLR 59, CA.
[42] In *Minja Properties Ltd* v *Cussins Property Group plc* [1998] 2 EGLR 52 a tenant had to pay for the cost of replacing corroded window-frames with new aluminium frames of a slightly better design, as prudent work. See also *Wandsworth London Borough Council* v *Griffin* [2000] 2 EGLR 105.
[43] *Lurcott* v *Wakely and Wheeler* [1911] 1 KB 905 at 924 (Buckley LJ), the "leading case" according to Harman LJ in *Brew Bros Ltd* v *Snax (Ross) Ltd* [1970] 1 QB 612 at 632.

notoriously difficult to draw and is dealt with for this reason in a separate part of this Chapter. However, it is worth noting here that the limitation so imposed on the scope of a lessee's obligation to repair recognises the fact that he has only a finite interest in the premises.

5. A contrast may be made between the position of a long lessee and a short-term tenant. In the case of the latter, at least where he holds a residential lease, one would not expect him to have to undertake any onerous repairs at all. This is recognised implicity by the fact that section 11 of the Landlord and Tenant Act 1985 imposes on landlords a statutory repairing covenant with regard to the structure and exterior of residential tenancies for less than seven years. Indeed in some American jurisdictions the tenant is treated, so it appears, as in the same position with respect to a short lease of premises used as his temporary home as any consumer of a durable product. As a result he benefits from a warranty of habitability, which imposes minimum standards of fitness on the landlord, and only requires him to undertake to keep the premises in a clean and tidy state and to replace breakages.[44] Contrast the position of a long lessee – whether he has a lease of residential or business premises. On any view he has an estate in land and, depending on the length of the unexpired residue of the lease, a valuable asset. In this case there is, as already explained, no reason why the courts should not interpret literally a lessee's obligation in, say, a building lease, not only to repair but also where necessary to renew and rebuild the premises. By contrast, it would be only in the face of the clearest language that a short business lessee would be required to rebuild as opposed to repair demised premises.[45]

Proof of disrepair

Where the landlord claims that the tenant has broken a repairing obligation, the onus is on him to prove the fact of disrepair. This may be achieved by service on the tenant of a dilapidations schedule at the end of the lease, if that is the time in issue, with full

---

[44] See eg *Javins v First National Realty Corpn* (1970) 428 F 2d 1071 at 1074; *Green v Superior Court of City and County of San Francisco* (1974) 111 Cal Rptr 704.
[45] *Lister v Lane and Nesham* [1893] 2 QB 212, CA; also cases discussed below.

costings and apparently also a full summary of the claim.[46] In one "remarkable" case, however, where according to Harman J,[47] intellect was perhaps allowed to take precedence over any other process, the Court of Appeal ruled that the tenant of a 1960s constructed office building did not have to pay for the cost of remedial works to a basement, which had in the past let in water owing to its inherently defective construction.[48] At the time the landlord made its claim, the basement was dry and seemingly the materials (porous concrete) of which it was made had been undamaged by the previous water ingress. The fact that the basement would be unusable and unfit for any purpose when flooded in this way would not suffice to trigger repairing liability in the absence of physical damage, any more than a condition of persistent and serious condensation in a house would require of itself a landlord to replace metal frame windows which were not damaged with new wooden framed windows which would "sweat" less or not at all.[49] "As a matter of ordinary usage of English", it was said "that which requires repair is in a condition worse than it was at some earlier time".[50] Thus an express covenant to repair by a lessee cannot require him to make the premises fit for any particular use if they are physically undamaged. Even if the law is reformed by introducing a default covenant, as suggested by the Law Commission,[51] that covenant (which could in any case be contracted out of by the parties) would apply to the landlord and it would not extend to an obligation in relation to commercial premises to make them fit for any particular use. In any case, if it is sought to impose such a covenant on a lessee, this must be done expressly.

Repair does not require improvements

1. "Repair" involves making good a damaged article "so as to leave the subject so far as possible as though it had not been

---

[46] See Robinson, "Reforms in Practice" *Estates Gazette* September 11 1998, referring to the recent changes in practice to civil procedure rules.
[47] In *Minja Properties Ltd* v *Cussins Property Group plc* [1998] 2 EGLR 52 at 55F.
[48] *Post Office* v *Aquarius Properties Ltd* [1987] 1 EGLR 40.
[49] *Quick* v *Taff-Ely Borough Council* [1985] 2 EGLR 50, CA.
[50] *Ibid* at 53J (Dillon LJ).
[51] Law Com No 238 (1996), para 7.10.

damaged".[52] It also involves the principle that the tenant is not required completely to replace the premises. The tenant is not required by a covenant to keep in repair or to repair to improve the premises. There is difficulty in defining what is an "improvement" as opposed to a "repair", as has been judicially recognised.[53] Where a landlord carried out a modernisation programme to a house held on a short residential tenancy, which involved among other things replacing a flat roof with a pitched roof, as well as removing the front and rear elevations of the house, renovating its water dispersal systems and installing new inner blocks, the Court of Appeal held that this work went beyond the notion of "repair".[54] The work had increased the market value of the house by almost double and also prolonged its estimated life by some 30%. Three tests were discerned with a view to helping to decide whether work was repair or improvement. These could apply separately or together.

2. According to Mustill LJ the tests were: (1) whether the alterations went to the whole or substantially the whole of the structure or only to a subsidiary part of it; (2) whether the effect of the alterations was to produce a building of a wholly different character to that let; and (3) what was the cost of the works in relation to the previous value of the building, and what was their effect on the value and lifespan of the building.
3. Where a landlord sought to compel tenants of a mature industrial building to pay for the cost of extensive works designed to eliminate rusting in the steel frame of the building, the High Court held that the tenants would not be liable to pay for works on that scale, since the condition ante-dated the commencement of the tenants' leases, and it was from the date of the lease that a repairing obligation ran.[55]
4. If the tenant is being asked to replace an existing item with one of a better design, when it is possible for the work satisfactorily to be executed by using the equivalent materials to those which have worn out or fallen into disrepair, then such work would lie outside the scope of a covenant to repair and amount to an

---

[52] *Calthorpe v McOscar, supra* at 734 (Atkin LJ).
[53] *Morcom v Campbell-Johnson* [1956] 1 QB 106 at 114 (Denning LJ).
[54] *McDougall v Easington District Council* [1989] 1 EGLR 93, CA.
[55] *Plough Investments Ltd v Manchester City Council* [1989] 1 EGLR 244.

"improvement". For example, if the tenant were to be invited to replace worn-out night-storage heaters with an underfloor heating system at greater expense, this would not be a "repair" but an "improvement". It was also said in the context of a landlord seeking to obtain a rent increase under the defunct controlled tenancies regime, on account of improvements, that where he replaced certain installations, the landlord had not carried out improvements since "there was no provision of anything new for [the tenants'] benefit, but only the replacement of old parts by a modern equivalent ... that does not amount to an improvement".[56] If a new central heating system had been installed where there had been none before, that would presumably consitute an improvement outside a tenant's covenant to repair.[57]

5. The articificiality of the line between repairs and improvements appears from the fact that on the other hand, replacing a defunct damp-proof course with a new one was held a repair,[58] whereas installing into an old building a damp-proof course where there was none before would amount to an improvement.[59] However, the tenant cannot limit his obligation to repair to exact replacement of like with like where the design of the original item was faulty. As a result, it was held that a lessee of a block of maisonettes had to pay for the cost of installing new expansion joints which had been omitted from the design of the building, so as to secure the outside cladding of the premises, even though the building was thereby improved.[60]

Lessee's liability to cure inherent defects

While a lessee is not obliged under a repairing covenant to improve the basic design of the demised premises, he may still have to cure the effects of an inherent design fault which had been present in the property as from the time it was let, if the work overall is on a scale

---

[56] *Morcom* v *Campbell-Johnson, supra* at 116 (Denning LJ).
[57] *Pearlman* v *Keepers and Governors of Harrow School* [1979] 1 All ER 365, CA.
[58] *Elmcroft Developments Ltd* v *Tankersley-Sawyer* [1984] 1 EGLR 47, CA.
[59] *Pembery* v *Lamdin* [1940] 2 All ER 434, CA; also *Wainwright* v *Leeds City Council* [1984] 1 EGLR 67, CA.
[60] *Ravenseft Properties Ltd* v *Davstone (Holdings) Ltd* [1980] QB 12.

that it is not complete or near-complete rebuilding of the whole premises. Accordingly, if there is damage to the premises caused by an unsuspected inherent defect, "it may be necessary to cure the defect, and thus to some extent improve without wholly renewing the property as the only practicable way of making good the damage to the subject-matter of the repairing covenant".[61] Thus the High Court ruled that it might well be necessary for lessees to pay for more limited remedial works than those proposed by the landlords, to cure immediate physical damage, such as bricks falling off the exterior of the building.[62] Provided physical damage is caused by the inherent defect, a lessee may find himself having to pay for work which will cure it, as occurred to a lessee of a block of maisonettes, who had undertaken an unqualified general obligation to repair.[63] The cost of the remedial work (the insertion of expansion joints omitted when the building was put up) was small when compared to that of total rebuilding. The cost of curing the inherent defect was one-tenth of the total repair bill. The work fell within the lessee's covenant. By contrast, a lessee was held not liable to pay for the replacement of the defectively-built foundations of a house.[64]

If an inherent defect is not at the time in question causing any physical damage to the premises, the tenant cannot be made to carry out or pay for remedial work to remove it.[65] The tenant also has a defence if he shows that the extent of any remedial work to a patent inherent defect lies outside the scope of his repairing obligation.[66] The risk of liability to cure inherent defects has been said to be greater in the case of new buildings, owing to the fact the court compares the state of the property as the parties contemplated that it would be in when the letting began and its actual state as damaged by the defect, so inviting a comparison with properly designed and constructed buildings.[67]

One method of avoiding or reducing the risk of liability to pay for work to cure inherent defects is for a potential lessee to procure

---

[61] *Quick* v *Taff-Ely Borough Council* [1985] 2 EGLR 50 at 52H (Dillon LJ).
[62] *Plough Investments Ltd* v *Manchester City Council* [1989] 1 EGLR 244.
[63] *Ravenseft Properties Ltd* v *Davstone (Holdings) Ltd, supra.*
[64] *Collins* v *Flynn* [1963] 2 All ER 1068.
[65] *Post Office* v *Aquarius Properties Ltd* [1987] 1 EGLR 40, CA.
[66] *Lister* v *Lane and Nesham* [1893] 2 QB 212, CA.
[67] Lewison, p 188.

an express term in the lease which limits or excludes any liability to repair if the root cause of the damage is a latent inherent defect at the time he inspected the premises prior to taking a lease. One precedent[68] suggests that the tenant would have an exemption from liability if the landlord would be able to claim for it against any third party who built or designed the premises. With regard to contracts entered into after the commencement of recent legislation, where a landlord has a right of action against contractors or other third parties owing to the poor design of a new building, and in terms contracts with them for the benefit of the tenant of the premises, the latter can sue the third parties direct.[69]

One further suggestion is to attempt to exclude liability of a lessee to pay for works attributable to an inherent defect.[70] A different method might be to exclude all liability for remedial works attributable to an inherent defect for a specific period after the lease begins. This latter limitation of liability recognises that even with a new building, as time runs on, the effects attributable to an inherent defect alone become presumably harder to isolate. A further method of mitigation of the real risks of liability for inherent defects might be for the tenant of a new building to try to extract a covenant from the landlord himself to remedy any defect in the building.[71]

C – Covenants to put, keep and leave in repair

Some leases do not merely impose on the tenant an obligation to "repair" – they qualify it in various ways, with the aim of imposing a stricter degree of obligation on the lessee.

Covenant to put in repair

A specific covenant by the lessee to "put" premises in repair might be used where the premises are being let in a poor state of repair, the tenant being thus obliged to put them into a suitable state of

---

[68] *Precedents for the Conveyancer*, No 5–66.
[69] Contracts (Rights of Third Parties) Act 1999, s 1. The Act, s1 of which can be contracted out of, applies to contracts entered into as from six months from November 11 1999. See further Chapter 4 of this book.
[70] Luxton and Wilkie, p 251.
[71] Ross, I, p 355.

repair, having regard to their age, character and locality. The standard of repair, where appropriate, will be that required for the purpose for which the premises have been let[72] – perhaps in accordance with an agreed dilapidations schedule. However, since a covenant to "keep in repair" by necessary implication requires the tenant to remedy initial dilapidations,[73] it would presumably only be in a case of a lessee's covenant to repair that a further undertaking to "put" in repair would add anything to his obligation.

Covenant to "keep in repair"

Where a tenant has expressly undertaken to keep premises in repair, this particular form of words emphasises the strict nature of the tenant's obligation, when compared to a covenant simply "to repair". He is under an obligation to see to it that any initial dilapidations in the premises, the remedying of which falls within the scope of a repair, are put right. If, during the term of the lease, part of the premises becomes phyically damaged, this state of affairs must be remedied by the lessee as soon as the damage has arisen. By analogy with the position of the landlord, where this principle has been established, the lessee's liability in damages would presumably run as from the date the damage to the premises was shown to have arisen.[74] Liability is judged objectively. It is not dependent on fault by the tenant being shown as a condition precedent to his liability to carry out work or pay damages. A tenant who fails to carry out regular inspections may find that he is liable to carry out or pay for remedial work resulting from damage to the premises which can be traced back originally to a fortuitous event, such as a blockage in an external downpipe caused by a dead pigeon,[75] or leaves blocking gutters which have not been regularly cleared. At the risk of repetition, it will be appreciated that the tenant could be at risk in accepting a covenant to "keep in repair" or to "keep in repair and good condition" unless the landlord has made good any physical damage revealed to the tenant, as by his surveyor, prior to his taking the lease, or the tenant

---

[72] *Belcher* v *M'Intosh* (1839) 2 M&R 186.
[73] *Proudfoot* v *Hart* (1890) 25 QBD 42, CA.
[74] *British Telecom plc* v *Sun Life Assurance Society plc* [1995] 2 EGLR 44; *Passley* v *London Borough of Wandsworth* (1996) 30 HLR 165, CA.
[75] *Bishop* v *Consolidated London Properties Ltd* (1933) 102 LJKB 257.

is confident that there is only a low risk of inherent defects[76] manifesting themselves during the lease, or again that he has obtained an exclusion of liability in respect of such defects during the lease. If the landlord shows that the defect or design fault is producing physical damage at the date of the action, the tenant must pay for the remedial work, on principles already discussed, if his repairing obligation is unqualified and if the cost of the work and other relevant factors point, as a matter of fact and degree, to the work being a repair rather than total renewal to the premises.

Covenant to leave in repair

Sometimes a tenant is expressly obliged to leave the premises in repair. If this undertaking comes in addition to a covenant to "keep in repair and good condition throughout the term" for example, it might nowadays seem unnecessary, in view of the strictness of the latter obligation. However, if the sole repairing obligation of a lessee is to leave in repair, the landlord must wait until the end of the lease before bringing any action for breach of the lessee's repairing covenant. For this reason it may be that this particular arrangement is only likely to be found in the case of short residential tenancies, where the condition of the premises is not likely to be much different at lease end from its condition when the lease began.

D – Repair, subordinate renewal and rebuilding

Only in the face of clear language is a lessee obliged to pay for or to carry out total renewal or rebuilding. As was seen in Chapter 4, however, there are circumstances in which a lessor undertakes to carry out major renewals in return for payment by the lessee of the cost.

Meaning of repair and subordinate renewal

Atkin LJ described "repair" as meaning the making good of damage so as to leave the subject as far as possible as though it had

---

[76] Including eg failure of tyings, steel frame rusting, failure of concrete used in the construction of the premises, failure of cavity walls, flat roof failures (which – see Hollis and Gibson, p 74f – may be hard to detect), and defects in chimney stacks (*ibid*, p 82).

not been damaged.[77] As a result, a tenant whose repairing obligation is unqualified may well find himself, provided the work is part of a repair, liable to pay for the cost of inserting a feature not present before, such as new expension joints to cladding, which had been omitted from the orginal design of the building.[78] A repairing obligation includes, as part of the process of replacing physically damaged items, replacement of subordinate parts of the premises.[79] If that work involves eliminating an inherent defect, such as the missing expansion joints just mentioned, then the tenant must carry out the work as a whole even if he is to some extent improving the condition of the premises.[80] Subordinate renewal means the reconstruction of part of the premises, or even a substantial part of the premises, but not the reconstruction of the entirety or substantially the whole of the demised premises.[81] The line between subordinate renewal, which is within a lessees' repairing obligation, and total or near total rebuilding, which is not, is hard to draw and is largely one of fact and degree. The difference between the concepts of repair and renewal is best shown by examples.

1. A tenant held an old house under a 28-year lease, and was subject to a strong general covenant to repair ("well and substantially" to "repair and keep in thorough repair and condition"). The tenant was held liable to pay the costs incurred by his landlord in demolishing an unsafe flank wall of the house and rebuilding it with a damp-proof course, so as to conform to local building regulations then in force. The work was characterised as subordinate renewal because the whole premises were not being rebuilt as a whole.[82] Fletcher Moulton LJ said, however, that a tenant in a case such as this might, in the course of the lease, have to rebuild first one part and then others of the premises, so that by the time the lease ended he might have replaced the whole building in stages. At no time would any individual item of work be any other than subordinate renewal.

---

[77] In *Calthorpe v McOscar* [1924] 1 KB 716 at 734.
[78] As in *Ravenseft Properties Ltd v Davstone (Holdings) Ltd* [1980] QB 12.
[79] *Lurcott v Wakely and Wheeler* [1911] 1 KB 905, CA.
[80] *Quick v Taff-Ely Borough Council* [1985] 2 EGLR 50, CA.
[81] *Lurcott v Wakely and Wheeler, supra.*
[82] *Lurcott v Wakely and Wheeler, supra.*

2. A commercial lessee was liable, under the guise of subordinate renewal, to pay for the complete replacement of a worn-out roof (even where the cost was not far short of two-thirds of the cost of total rebuilding of the premises).[83] A commercial lessee had to pay for the cost of the underpinning[84] of part of the foundations of commercial premises.[85] The whole premises did not have to be reconstructed.
3. The cost of replacing a tilting flank wall which had been undermined by seepage of water was held to lie on the renewal side of the line where evidence had been produced that to replace the item would cost the same as putting up a new building.[86]
4. Lessees holding a term for seven years were not bound to pay for the cost of rebuilding the old house which had to be demolished. The reason for the demolition was the failure of the foundations of the premises. These had been built on a mud-sill foundation: to have saved the house would have entailed underpinning the walls of the house through several feet of the mud to solid ground.[87] The lessees could only have been required to repair the house they took – a property with defective foundations. A covenant to repair does not include a covenant to improve the nature of the property as a whole.
5. The High Court applied this decision to a 14-year lessee of a house which had a "jerry-built" structure built on to it at a later date, described as a "utility room". The room had reached the end of its useful or effective life. The lessee was not required to rebuild the room. The cost of rebuilding was one-third of the cost of rebuilding the premises as a whole.[88] The lessee could not be required to pay for improvements. However in this case the High Court appears to have treated the "utility room" as a

---

[83] *Elite Investments Ltd v TI Bainbridge Silencers Ltd* [1986] 2 EGLR 43.
[84] A method stigmatised by Melville and Gordon, p 122 as "very expensive indeed".
[85] *Rich Investments Ltd v Camgate Litho Ltd* [1988] EGCS 132; also *Alexander v Lambeth London Borough Council* [2000] 2 CL 386.
[86] *Brew Bros Ltd v Snax (Ross) Ltd* [1970] 1 QB 612, CA.
[87] *Lister v Lane and Nesham* [1893] 2 QB 212, CA; also *Weatherhead v Deka New Zealand Ltd* [2000] 1 NZLR 23.
[88] *Halliard Property Co Ltd v Nicholas Clarke Investments Ltd* [1984] 1 EGLR 45.

separate structure from the rest of the premises, since it was added on at a later date.

6. A lessee who takes premises with a basement which lacks a damp-proof course cannot be required to insert a damp-proof course for the benefit of the landlord,[89] whether or not the cost of the work would be treated as a repair if it related to replacing some item in the premises which had become in need of replacement, such as a door. The Court of Appeal recently ruled that a Rent-Act protected tenant, who was under a widely worded repairing obligation, was not liable to pay for the insertion of a damp-proof course at a cost of some £15,000 into a house, suffering from rising damp, built in 1841 with shallow foundations and with no damp-proof course. Regard was paid to the poor condition of the premises when let and to the limited interest of the tenant. The principle was invoked that to have required the tenant to do this work would have involved giving the landlord back a different thing to that demised.[90]

7. A tenant who had a timber house was not required to alter the basic construction of the house and return to his landlord a stone house.[91] A tenant who held premises with a latent inherent defect in the form of a defective "kicker" joint in the basement did not have to pay for asphalt tanking to make sure that at some uncertain future date the basement would be protected from water invasion.[92] In an Australian case, it was held that a tenant did not have to pay for the cost of curing a latent inherent defect which in due course would lead to the collapse of the whole premises.[93] There had seemingly been no proof by the landlord of actual or imminent physical damage to the premises – if there had been, the landlord would have had to prove in addition that any remedial work was not total rebuilding (unless the lease imposed a clear obligation on the lessee to carry out rebuilding).

---

[89] *Pembery* v *Lamdin* [1940] 2 All ER 434, CA.
[90] *Eyre* v *McCracken* (2000) 80 P&CR 220, CA. A submission by the landlords that regard should be had to the statutory protection of the tenant was not accepted.
[91] *Gutterridge* v *Munyard* (1834) 1 M&R 334 at 336.
[92] *Post Office* v *Aquarius Properties Ltd* [1987] 1 EGLR 40, CA.
[93] *Clowes* v *Bentley Pty Ltd* [1970] WAR 24.

Equivalent replacement

Subject to the fact that an inherently defective item cannot very well be replaced with its exact equivalent, a damaged item must be replaced by its equivalent, if such is obtainable. We have already seen that where an underfloor heating system had become faulty because a number of circuits were broken, due to differential rates of expansion causing cables not fitted with flexible connectors to break, as well as to building subsidence, the tenant still had to replace the damaged parts of the underfloor heating system rather than replace it with night-storage heaters.[94] The Court of Appeal regarded maintenance of the existing heating system as possible and practicable on the given facts. The fact that extensive repairs would be needed was immaterial. It is not open to the tenant to select a method of curing physical damage merely on cost grounds, where the damaged item can be restored.[95] It was accepted that if the tenant had shown that an attempt to repair with like for like would be futile, then the covenant could have been complied with by substituting some other equivalent materials or system. An example of a futile repair might be that of replastering a wall afflicted with damp,[96] caused by the failure of a damp-proof course,[97] as opposed to replacing the damp-proof course itself. Sometimes the correct course of action will be suggested only by the facts – thus, by analogy with landlords' statutory implied covenants, a door may have reached the stage that it is so damp or otherwise defective that the only sensible course of action is to replace it as part of the obligation to keep the item in repair.[98]

---

[94] *Creska Ltd v Hammersmith and Fulham London Borough Council* [1998] 3 EGLR 35, CA.

[95] The implications of this reasoning may be severe. A tenant faced with roofing repairs involving complete stripping and replacement of slates, opting for a quality of slates with a significantly lower life-duration than those being replaced (as to expected life duration of different roofing materials see IS Seeley, *Building Surveys*, 1985, p 69) might be in breach of covenant. If a practical, reasonable surveyor would support more than one method, the tenant could seemingly choose: *Ultraworth Ltd v General Accident Fire & Life* [2000] 2 EGLR 115 at 118.

[96] In *Elmcroft Developments Ltd v Tankersley-Sawyer* [1984] 1 EGLR 47, a landlord was obliged to remedy a failure of a dpc by silicone injection not replastering the damp wall.

[97] As to various methods of treatment see eg BRE Digest 245 "Rising Damp in Walls: Diagnosis and Treatment"; Melville and Gordon, pp 257–258.

[98] As in *Stent v Monmouth District Council* [1987] 1 EGLR 59, CA.

Amount of repairs required

The amount of work required to comply with a repairing obligation is a question of fact. It is not possible for a tenant to subject his liability to any implied limitation that the benefit of the remedial work ought not to outlast the duration of his lease.[99] The question may arise in relation to roofing repairs, where the landlord may be pressing for wholesale stripping and replacement of the exisiting roof cover and the tenant for patch repairs. The issue is one of degree – as to what will suffice to protect the structure of the premises from rain penetration and resulting damp. In the context of the statutory implied repairing obligation, a landlord was held to have been in compliance with his obligation by making patch repairs to the roof of an old house, on the facts.[100] But if the evidence shows that the stage has been reached at which the roof cannot realistically be saved, or that patch work is not a practicable option,[101] then it would be open to the court to hold that the costs of full replacement of the roof fell on the tenant's shoulders as a repair.

Implied limits on restoration work

It seems that a tenant cannot not be required to carry out restoration of a feature if to do so would conflict with the building regulations in force at the time of the landlord's claim. A tenant was not required to rebuild a bay window save as he did – flush with the main walls – where to have restored the original design of the window would have been contrary to building regulations unless substantial extra support for the projecting structure was inserted, so improving the premises.[102]

The extent of a tenant's obligation to carry out works may well be limited where he has taken a lease of a mature building whose construction was by a method or materials which have subsequently shown themselves to be inadequate. The courts, as seen, have regard to the age of the premises and their condition when let. They are reluctant to impose on a lessee who is subject to an obligation to repair any implied requirement to improve the

---

[99] *Ladbroke Hotels Ltd* v *Sandhu* [1995] 2 EGLR 92.
[100] *Dame Margaret Hungerford Charity Trustees* v *Beazeley* [1993] 2 EGLR 143.
[101] As to which see Hollis and Gibson, p 80 (pitched roofs).
[102] *Wright* v *Lawson* (1903) 19 TLR 510, CA.

state or construction of the building, when compared to its state when demised. Thus, the High Court refused to declare that a 20-year lessee of a 55-year-old industrial building was liable to defray the cost of work which would cure a rusting condition in the main frame of the premises.[103] The condition existed at the date of the lease. The lessees were held to be bound to cure damage to the external brickwork but not to remove the root cause of the damage – on the ground that the condition was present in the building when it was let. The work required by the landlord was so extensive as to go outside the lessees' repairing obligations. A similar approach was adopted by the New Zealand Court of Appeal.[104] A restaurant formed part of a 66-year-old retail building let to tenants on a 40-year lease. The materials used to make the structural concrete bands in the restaurant consisted of aggregates which were liable to break down and crumble. Although the lessees were subject to a widely-worded repairing obligation, they were held not liable to pay for the heavy cost of renewal to the restaurant. To have decided otherwise would have provided the landlords with a windfall in the form of a substantially and permanently improved building, when compared to the premises as let. In the words of Windeyer J, in cases such as these, the tenant is not required to put the premises into "mint condition".[105]

Express exception for fair wear and tear

Some leases contain an express exception for fair wear and tear, the aim being to prevent the tenant being liable for work caused merely by the effects of ageing and weathering. If the tenant can show that the effect of these factors is greater on the premises than it would be on new premises, the exception may relieve him from liability, since he is not liable to hand back to the landlord, at the end of the lease, a brand-new building in appropriate condition, if he took an old or mature building. In addition, a fair wear and tear exception prevents a tenant being held liable for the normal and reasonable use of the premises for the purposes for which they were let.[106] An express fair wear and tear exception is limited in scope.

---

[103] In *Plough Investments Ltd* v *Manchester City Council* [1989] 1 EGLR 244.
[104] In *Weatherhead* v *Deka New Zealand Ltd* [2000] 1 NZLR 23.
[105] In *Alcatel Australia Ltd* v *Scarcella* (1998) 44 NSWLR 349 at 354.
[106] *Terrell* v *Murray* (1901) 17 TLR 570; *Miller* v *Burt* (1918) 63 SJ 117.

1. It does not apply to any damage caused by an abnormal or exceptional event, such as fire, lightning or earthquake.
2. A tenant's covenant to keep in repair implies an obligation to leave premises in repair, and hence, in the same condition as when let, allowing for the passage of time. Though a tenant is assumed, in the case of a long lease, to have carried out repairs as and where needed,[107] a fair wear and tear exception emphasises that the tenant is not bound to hand back to the landlord a building of a different character, age or condition, to that leased, and allows for the fact that the tenant or his predecessor may have been in possession for a long period, during which ageing as well as disrepair has taken place. In order to narrow the focus of disputes as to the condition of the premises when let, a jointly agreed schedule of condition may be drawn up at the time the lease is granted.[108]
3. The tenant cannot use a fair wear and tear exception to avoid liability for physical damage caused by disrepair, and it covers direct and not consequential loss. Thus, if a slate falls off the roof because its tingle is worn out, while the tenant could claim that he did not have to replace the worn-out slate owing to the fair wear and tear exception, the House of Lords have held that he would be liable for consequential damage to the roof rafters from water ingress.[109] Thus the tenant's easiest course of action might be to replace the slate concerned.

E – Redecoration as part of repairing obligations

Liability arising under covenants to paint at regular intervals

With a view to protecting wood and other surfaces from decay and damage, leases may well contain a covenant by the tenant (or in the case of a multi-occupied building the landlord) to carry out regular repainting and redecoration at specified intervals.[110] Thus, in the case of a long lease the tenant might have to repaint the exterior at

---

[107] *Calthorpe v McOscar* [1924] 1 KB 716, CA.
[108] See Seeley, pp 164 – 165.
[109] *Regis Property Co v Dudley* [1959] AC 370, approving *Haskell v Marlow* [1928] 2 KB 45.
[110] See eg *Precedents for the Conveyancer* 5–8, cl 3(6)(b); 5–10 Sched 4 cl 4(2); 5–31, cl 5(5)(d) and 5–52.

three- or five-yearly intervals, or, if intervals are not specified, as and where required. In the case of interior decorations, seven-yearly intervals may well be encountered. Whereas the need for regular short intervals in the case of external repainting is obvious in view of the relatively short life of paint coats,[111] it is not apparent that internal decorations, if properly carried out in the first place and suitably maintained subsequently, will necessarily need to be carried out after a seven-year interval, which period has been characterised as "often excessive".[112]

Perhaps owing to these considerations,[113] section 147 of the Law of Property Act 1925 provides special relief where a landlord is attempting to forfeit a lease for breach of covenant to carry out internal decorative repairs. If the court is satisfied that the landlord's forfeiture notice in relation to such repairs is unreasonable, it has a discretion wholly or partly to relieve the tenant from liability for these repairs. However, section 147 does not apply to a covenant by a lessee to yield up the premises in a specified state of repair at the end of the term.

In the absence of physical damage to internal decorations the tenant is seemingly not liable, unless the lease provides to the contrary, to carry them out even if the relevant interval has passed, and is not liable to carry out interal decorations if he has repainted at the specified intervals and leaves without repainting, if all the paintwork requires is cleaning.[114] Where there is an obligation to repaint which arises in a specific year, then in the absence of some express provision to the contrary, liability begins at the start of the year.[115] If, within a specified interval, the woodwork becomes exposed owing to failure of the paint film, the tenant should carry out remedial work – at least to the affected part – as part of his

---

[111] As to which see eg Melville and Gordon, pp 198–199; also Hollis and Gibson, p 151.
[112] Melville and Gordon, p 199.
[113] And also to prevent unconscionable pressure being brought to bear on a tenant to buy up the reversion (*Wolstenholme and Cherry's Conveyancing Statutes* 13th edn Vol 1 p 273).
[114] *Scales v Lawrence* (1860) 2 F&F 289; it was held in *Perry v Chotzner* (1893) 9 TLR 488 that trivial matters such as holes left by picture-hooks did not need to be made good in between specified decoration intervals.
[115] *Kirklinton v Wood* [1917] 1 KB 332 (obligation to paint arose on 1 January: tenant in breach as from that date).

general obligation to repair. The effect of having specified intervals may limit the obligation to repaint. In the absence of some express provision, such as imposing liability if the lease is sooner determined, if a lessee terminates his lease by notice at the end of, for example, year five, he would not be bound to carry out painting if the obligation to do so arose only at year seven.[116]

Liability to paint under a general covenant to repair

A lease may not impose on the tenant any specific obligation to paint or decorate during the term, but may require the premises to be kept in repair, or to be left in repair at the end of the lease. Where this is the case, a tenant is only liable to carry out such painting and decorating as will protect the wood and other surfaces such as plasterwork from physical damage.[117] Thus the tenant of a dwelling-house holding under a short lease who was under a covenant to keep and leave the premises in good tenantable repair was required to paint and decorate any item, at least before quitting, where a reasonably-minded incoming new tenant would expect that work to be done[118] – assuming that the surfaces concerned were exposed, requiring renewal of the painting or decorations.[119] The standard of any redecoration would seem to vary with the district where the premises are situated.

F – Standard of repair

The terms of the lease may specify the standard of repairs required from the lessee, but if they are silent, the issue is resolved under the general law. There is a difference between the principles applying to short tenancies and long leases.

---

[116] *Dickinson* v *St Aubyn* [1944] 1 All ER 370.
[117] *Monk* v *Noyes* (1824) 1 C&P 265.
[118] *Proudfoot* v *Hart* (1890) 25 QBD 42, CA.
[119] *Crawford* v *Newton* (1886) 36 WR 54 (internal repainting needed under general covenant to repair only to prevent decay even if, as there, no internal decorations have been carried out for some 17 years).

Short tenancies

If a short tenant is under an obligation to keep and deliver up the premises in "good tenantable repair,"[120] the common law holds that the governing requirements are those of a reasonably-minded incoming tenant.[121] The outgoing tenant is bound to keep the premises in such repair as, having regard to their age, character and locality, would make them reasonably fit for the occupation of a reasonably-minded tenant of the class likely to take them. Thus if the premises are situated in an expensive locality, the standard of repairs required of an outgoing tenant is in principle likely to be higher than those located in a more modest or a run-down area. However, if the requirements of reasonably-minded incoming tenants have fallen since the lease was granted to the outgoing tenant, to that extent he has a defence and is presumably only bound to repair to the new, lower, standards. In the short time of a short-term tenancy such a limitation is presumed not to be very likely to arise. We have already seen that if paintwork or other surfaces requiring decoration are damaged, then the outgoing tenant is liable to cure this damage – the extent of the cure being a question of fact and in accordance with the subjective standard required by the common law. If the wallpaper, for instance, was adequate for incoming tenants, it would not need replacing. If it was peeling off the walls, then clearly replacement with new paper would be called for. The quality of the paint to be used in any redecoration required need not be better than that used when the tenant took the tenancy – provided that the paint being replaced complied with the covenant to repair. Similarly, an old floor need not be replaced except to the extent that it may have become rotten – and so unacceptable to any reasonable incoming tenant. In the context of a business lease, it was held that incomplete fire protection equipment was not in good tentantable condition and so not acceptable to an incoming tenant.[122]

---

[120] The effect of s 11 of the Landlord and Tenant Act 1985 where applicable on such an obligation is to limit it to interior parts of the demised premises.
[121] *Proudfoot* v *Hart, supra* – applying as much to functional equipment such as an air-conditioning plant as to decorative items: *Land Securities plc* v *Westminster City Council (No 2)* [1995] 1 EGLR 245 at 255L.
[122] *Shortlands Investments Ltd* v *Cargill* [1995] 1 EGLR 51.

Long leases

If the tenant holds a long lease, the Court of Appeal laid down the following strict standard where the lease makes no express provisions.[123] The tenant must undertake such repairs as may be required to put the premises into the state of repair and condition as they would have been in, if they had been managed by a reasonably-minded owner. Full regard is paid to the age of the premises, as well as their location, and the class of likely occupying tenants. Equally it is to be assumed that the property has been kept up in such a way that only an average amount of annual repairs are required. The standard of repairs is thus not exclusively governed by the subjective requirements of a notional reasonably-minded incoming tenant. The Court of Appeal therefore rejected an argument by a lessee holding three 95-year leases of houses, that he was only bound to pay damages assessed on the basis of repairs needed to satisfy incoming tenants of the kind likely at the end of the lease to take tenancies of the premises – these requirements had fallen considerably since the date of granting the leases in the early part of the nineteenth century. The character of remedial work required in applying this standard is presumably one of fact and it is arguable be that if the use of the premises has altered during the life of the lease, their current use determines the relevant stardard of work.

G – Liability to repair or put up additional buildings

Some general aspects

An unqualified lessee's general covenant to repair extends to all buildings erected by him during the term of the lease.[124] The extent of the obligation is, however, a matter of construction of the particular covenant. Thus a tenant who covenanted to repair the "buildings demised" was held liable only to repair the specific buildings let.[125] If the landlord wishes to extend a covenant to repair to include buildings to be put up during the lease, therefore, clear language is required – if a covenant is too specific, it may not apply to a particular building put up during the lease term by the

---

[123] *Calthorpe* v *McOscar* [1924] 1 KB 716, CA.
[124] *Hudson* v *Williams* (1878) 39 LT 632.
[125] *Doe d Worcester Trustees* v *Rowlands* (1841) 9 C&P 734.

tenant.[126] By contrast, where a long lessee covenanted to build a certain number of houses and then to keep in repair "the said premises" and built six houses on the land concerned, his covenant to repair was held, as a matter of construction, to extend to all houses built, in view of its generality. It was envisaged that houses would be put up by the lessee, rather than his just repairing buildings already on the land.[127] Since chattels forming part and parcel of the land (or fixtures) form part of any relevant buildings, an obligation to repair is taken, unless the lease provides to the contrary, to extend to landlords' fixtures attached by the tenant during the lease.[128]

Breaches of covenant to build

If a building lessee is under covenant to erect certain buildings by a specified date, he is not impliedly required to put up any buildings before that date.[129] Equally, a lessee who fails to complete the buildings by the specified date incurs a forfeiture of his lease.[130] If a lessee has broken a covenant to reconstruct premises by a certain date, such a breach is capable of being remedied for forfeiture purposes, albeit that the relevant date has expired.[131]

If the tenant finds that for the time being he cannot comply with a covenant to build, as where planning restrictions or war-time regulations prevent building, he has a defence of temporary impossibility of performance to any action by his landlord for breach of covenant: the covenant revives if the temporary impossibility ceases. Thus, a lessee who was unable owing to the listing of a building to demolish it, had a lawful excuse for not complying with his covenant – the listed building consent rules having supervened since the grant of the lease.[132] If the demised

---

[126] As in *Smith v Mills* (1899) 16 TLR 59.
[127] *Field v Curnick* [1926] 2 KB 374.
[128] If the tenant installs trade fixtures and renews his lease, his repairing obligation is taken, in the absence of contrary language, to extend to these fixtures: *Thresher v East London Waterworks Co* (1824) 2 B&C 608.
[129] *John Lewis Properties plc v Viscount Chelsea* [1993] 2 EGLR 77.
[130] *Bennett v Herring* (1857) 3 CB (NS) 370.
[131] *Expert Clothing Service & Sales Ltd v Hillgate House Ltd* [1985] 2 EGLR 85, CA.
[132] *John Lewis Properties plc v Viscount Chelsea, supra.*

land were to be destroyed by some natural disaster such as an earthquake or sea invasion, the tenant would seemingly have a complete defence to liability based on frustration of the lease.[133]

III – Covenants against structural alterations to the premises

A – Absolute prohibitions

Some leases contain a covenant against the making of structural or other alterations by the lessee. The covenant may take the form of an absolute prohibition on the making of alterations,[134] or it may be qualified or fully qualified. The expression "alteration" is limited to anything which significantly alters the form and construction of the building.[135] A lessee who attached a large clock to the wall of the premises (which item was taken as convenient to the conduct of the lessee's business) had not carried out an "alteration".[136] Likewise, a lessee who puts up advertisements, provided these are incidental to the purpose of his business, would not in principle be within the mischief of a covenant against alterations, provided no lasting damage is done to the premises.[137]

If the tenant breaks an absolute covenant against the making of structural alterations, the landlord may seek a mandatory injunction compelling him to restore the premises to their previous condition. Where there had been an "almost subversive" unauthorised alteration – a reduction to the height of a parapet wall by some nine to twelve inches – the alteration could not be deemed minor: a mandatory injuction was awarded against the tenant to compel him to restore the damage.[138] If the landlord wishes to

---

[133] See *National Carriers Ltd v Panalpina (Northern) Ltd* [1981] AC 675, HL (frustration not shown on facts).
[134] See eg *Precedents for the Conveyancer* 5–83 Sched 4 Part II para 1(7).
[135] In the case of a covenant not to carry out external alterations, holes in the wall would amount to a breach: *London County Council v Hutter* [1925] Ch 626; see further Crabb, *Leases: Covenants and Consents* (1991) p 150.
[136] *Bickmore v Dimmer* [1903] 1 Ch 158. Contrast *London County Council v Hutter, supra*: a lessee who put up T irons and brackets which were not reasonably incidental to his trade was ordered to remove these and to make good the damage to the stonework of the premises.
[137] *Joseph v London County Council* (1914) 111 LT 276 (easily removable frame, where covenant referred only to alterations of the fabric).
[138] *Viscount Chelsea v Muscatt* [1990] 2 EGLR 48, CA.

forfeit a lease for breach of this particular covenant, it is not essential that he has precise details, provided he has clear general evidence, as from photographs taken from other premises, of work which is in breach of covenant.[139]

B – Statutory relief

Statute outlaws absolute prohibitions in the case of Rent Act 1977 protected and statutory tenancies.[140] Specific statutory provisions allow the county court to alter or modify a restrictive covenant which could impede structural or other alterations to premises let as a factory, or as offices, or with respect to fire precautions where a fire certificate has become compulsory.[141]

Statute allows a local authority or lessee to apply to the county court to vary an absolute prohibition on structural alterations to premises held on a long lease. The idea is to allow for the creation of smaller units.[142] One gateway to an order is to show that owing to changes in the character of the neighbourhood of the premises,[143] they cannot readily be let as a single dwelling-house but could readily be let for occupation if converted into one or more dwelling-houses. This provision was held not to apply to a scheme for division of adjoining terraced houses into flats, even if the flats might extend to the full width of the former houses.[144]

---

[139] *Iperion Investments Corporation* v *Broadwalk House Residents Ltd* [1992] 2 EGLR 235.
[140] Housing Act 1980, ss 81–83, as amended.
[141] Factories Act 1961, s 169; Offices, Shops and Railway Premises Act 1963, s 73; Fire Precautions Act 1971, s 28. Law of Property Act 1925, s 84 allows any person interested under a lease granted for a term of over 40 years or more to apply to the Lands Tribunal, once 25 years of the lease have expired, for the discharge or modification of an absolute prohibition on the making of structural alterations.
[142] Housing Act 1985, s 610.
[143] As to which see *Alliance Economic Investment Co* v *Berton* (1923) 92 LJKB 750.
[144] *Josephine Trust* v *Champagne* [1963] 2 QB 160, CA. The Law Commission (*Covenants Restricting Disposals, Alterations and Changes of User* (1985) No 141, para 9.15) recommended a statutory reversal of this decision.

Disability discrimination

A significant effect on covenants against alterations, whether absolute or fully qualified, is the result of Part III of the Disability Discrimination Act 1995, to be brought into force as from October 1 2004. The policy of Part III of this Act is to make sure, within some limits, that disabled persons, such as the blind or those confined to wheelchairs, are not prevented from gaining access to goods, facilties and services provided, in particular, by traders and other business persons. When section 21(2) of the Act comes into force, a lessee, as a "service provider" will be required to take reasonable steps to remove or alter any physical feature[145] which makes it "impossible or unreasonably difficult" to make use of the service in question. The lessee might, for example, have to remove the offending feature, or alter it so that it no longer has the prescribed effect. Thus a shop lessee might find himself having to widen doors or provide lifts, or permanent ramp access, depending on the circumstances.

Where the lessee is subject to an absolute or fully qualfied prohibition on the making of structural alterations in his lease, the 1995 Act provides that the statutory duties cannot be avoided by the mere existence of the covenant, whatever its form. Indeed, the lessee appears to be treated by section 27 of the 1995 Act as being entitled to carry out any alteration to the premises which he might have to make to be in conformity with his statutory duties. His exposure, as a result, as from the coming into force of Part III of the 1995 Act, to an action for breach of statutory duty by a disabled person claiming denial of access provides an incentive for a lessee who is within the Act (such as a business tenant) to apply to his immediate landlord in writing for consent to any necessary work. If he fails to apply, then the fact that his lease may prohibit the carrying out of the work will not be taken into account as a defence if the lessee faces an action by an aggrieved disabled person for breach of statutory duty.[146]

If the landlord refuses consent unreasonably or imposes conditions on his consent, the lessee[147] may refer the matter to the

---

[145] Defined in Disability Discrimination etc Regs 1999 No 1191, regs 2 and 3.
[146] Disability Discrimination Act 1995, Sched 4, para 5. The lease is in principle treated as providing for the lessee to apply in writing for consent (s27(2)(6)).
[147] And also any disabled person with an interest in the proposed alteration.

county court. It in turn must decide whether the landlord's refusal or any of his conditions, as the case may be, were unreasonable.[148] If the court rules against the landlord, it may so declare or by order authorise the lessee to make the alteration specified in its order. Regulations will in due course be made in which are set out the circumstances in which a landlord has withheld consent unreasonably or imposed unreasonable conditions when giving consent.[149] If the landlord imposes a condition requiring planning permission to be obtained where required, or allowing him to inspect the work when completed, it would seem that such conditions would in principle be upheld – but only time will tell.

The 1995 Act overrides the terms of alteration covenants whatever their form and irrespective of the length of the lease. The duty to make alterations is not absolute: it may be that some lessees could provide say a mail order catalogue to wheelchair users – giving them different, but equivalent access to the services in issue.[150] But in the case of, say, a restaurant, it appears that in the nature of the business, physical access would have to be constructed to allow mobility or visually impaired persons equivalent access to that enjoyed by others to the interior of the premises: these disabled people could not be said to have such equivalence if they could only make use of an outside part of the restaurant.

It is clear that the overriding principle of the 1995 Act is equality of access: the terms of standard business leases are overridden, a result seemingly aimed at by Parliament.[151] This consideration takes primacy over freedom of contract and of the economic interests of the lessee, who may find himself having to pay for the costs of alterations to the premises from which his landlord will ultimately benefit, as well as for the costs of an application for consent to the landlord.[152]

---

[148] 1995 Act, Sched 4, para 6.
[149] 1995 Act, Sched 4, para 8.
[150] Example from draft Code of Practice (2000) – it appears that a final version may be published in the spring of 2001 (drc.db.org).
[151] *Hansard* HL Vol 566 col 1016 and the commentary in *Current Law Statutes* on the 1995 Act.
[152] If the lessee is entitled to renew his lease under statute, he cannot claim compensation from the landlord under Part I of the Landlord and Tenant Act 1927.

C – Qualified and fully qualified prohibitions

A covenant against the making of improvements such as structural alterations[153] without the licence or consent of the landlord is deemed by section 19(2) of the Landlord and Tenant Act 1927 to be fully qualified – hence, the landlord cannot unreasonably withhold his consent. This deeming rule cannot be contracted out of in any lease – but it can be avoided by the landlord taking an absolute prohibition, in which case he can grant a licence for a particular alteration at his own discretion without review by the courts. The 1927 Act gives a landlord some protections.

1. The landlord may validly, as a condition of giving consent, require the lessee to pay him a reasonable sum in respect of any damage to or diminution in the value of the premises or any neighbouring premises of his, if such results from the "improvement". He may also require the payment of his legal and other expenses in connection with his consent (s 19(2) of the 1927 Act).
2. If a particular "improvement" does not add to the letting value of the holding, the landlord may require, as a condition of his licence or consent, an undertaking by the lessee to reinstate the premises to their previous condition at the end of the lease.

This particular right conferred on landlords may explain the fact that the test of whether a proposed structual alteration amounts to an "improvement" within the legislation is judged by whether the alterations would render the tenant's occupation of the demised premises more convenient and comfortable to him.

Thus, in a leading case,[154] the tenants of a shop held on a long lease wished to enlarge the premises by pulling down a rear wall and connecting the shop with adjoining land held from a different lessor. This would allow them to construct one large shop. The Court of Appeal held that the proposed alterations were "improvements" within section 19(2) of the 1927 Act. The tenant failed only because he had not proved that the landlord had

---

[153] A related covenant is an obligation not to cut any main walls or timbers without the landlord's prior licence or consent. On the construction of such a covenant, see *Hagee (London) Ltd* v *Co-operative Insurance Society* [1992] 1 EGLR 57.
[154] *FW Woolworth & Co* v *Lambert* [1937] Ch 37.

unreasonably withheld his consent. In subsequent proceedings, the landlord having refused consent without more, it was held that not only had the landlord unreasonably refused his consent but also had put it out of his own power, by not consenting, to protect himself by asking for either compensation or reinstatement.[155] Thus the landlord may only make use of his statutory rights when giving consent.

The tenant must ask the landlord for consent. If it is refused and the tenant takes proceedings for a declaration,[156] he must prove the landlord to have unreasonably refused consent. If the landlord gives no reason for a refusal of consent, however, then the onus of justifying his refusal is shifted to him.[157] The court might infer that a landlord who advances an obviously bad reason for his refusing consent has no valid grounds for his refusal operating on his mind at the time of his decision.[158]

A lessor may advance aesthetic, historical or personal grounds as the basis for a reasonable refusal of consent, but the issue is judged objectively. A landlord failed to convince the court that his objections to his tenant building a shed, extending a patio and building a retaining wall were well founded.[159] However, if the work proposed would trespass into the landlord's own adjoining premises, he is entitled reasonably to refuse his consent to the work as planned.[160] The landlord could also refuse consent reasonably if he showed that the work would trespass into his air-space.[161]

What a landlord cannot do is to refuse his consent to the work the tenant is proposing to carry out merely because there is only an improvement to the premises from the latter's point of view rather

---

[155] *Lambert v FW Woolworth & Co Ltd (No 2)* [1938] Ch 883, CA.
[156] As envisaged by Landlord and Tenant Act 1954, s 53; alternatively the tenant may carry out the work, with the prospect of relief against forfeiture if the landlord could not reasonably have refused consent.
[157] *Lambert v FW Woolworth & Co (No 2)* [1938] Ch 883 at 906 (Slesser LJ).
[158] *Lovelock v Margo* [1963] 2 QB 786 at 790, CA, decided prior to the reversal by Landlord and Tenant Act 1988, s 1 of the onus of proof of justifying the reasonableness of withholding consent to a proposed assignment.
[159] *McCulloch v Elsholz* (Unreported) January 16 1990.
[160] *Haines v Florensa* [1990] 1 EGLR 73, CA.
[161] *Davies v Yadegar* [1990] 1 EGLR 70 (where such was not proved by the landlord since the relevant air-space surrounding a proposed dormer window was demised to the lessee).

than from the lessor's. Thus, a landlord could not reasonably object to the tenant moving the position of a staircase in order to facilitate the business of his sublessee.[162] The same result was arrived at where the lessee proposed to make an aperture in the main wall in two places between the premises demised and neighbouring premises, so as to connect them up.[163] If the effect of the alterations carried out by the tenant is to add to the letting value of the premises, the tenant is entitled to claim compensation from the landlord at the end of the lease provided he complies with the strict conditions laid down by statute.[164]

IV – Rent review implications

A – Common law

If a tenant's improvement raises the letting value of the demised premises, it forms part of these premises. If a rent review clause in the lease requires a valuer to find an open market rental value for the demised premises, with no qualification on account of tenants' improvements, the effect of the latter must be reflected in the reviewed rent. It makes no difference that the work was carried out by the current tenant at his expense or by a predecessor in title of his under the same lease or even under a pre-lease occupation agreement.[165]

B – Disregard clauses

It may be that the ascertainment of rent payable at the date of a rent review[166] is required to an open market rental, but the rent review

---

[162] *Balls Bros Ltd* v *Sinclair* [1931] 2 Ch 325.
[163] *Lilley & Skinner Ltd* v *Crump* (1929) 73 SJ 366.
[164] Landlord and Tenant Act 1927 Part I – some landlords may offer themselves to do the work, so as to avoid having to pay compensation. If the landlord grants the tenant a licence to carry out the work, s 2(1)(b) of the 1927 Act may not rule out a later claim for compensation by the lessee, as it could be argued that no valuable consideration moves from the landlord (see *Precedents for the Conveyancer* 5–72 notes).
[165] *Ponsford* v *HMS Aerosols Ltd* [1979] AC 63, HL.
[166] Valuing the premises, in the absence of a contrary provision, as at the review date, so that a disregard clause may apply to buildings put up by sublessees of land undeveloped at the date of the lease: *Laura*

clause concerned may also require the valuer to exclude from this ascertainment any effect on the rent of tenants' improvements. Much turns, however, on the precise language of a disregard clause. One comprehensive-looking precedent[167] requires the disregard of any increase in the rental value of the demised premises which is attributable to any improvement to the premises, with consent where required,[168] otherwise than pursuant to an obligation to the landlord or his predecessors in title, where the work was carried out by the tenant, or sub-tenant, or their predecessors in title during the lease, or any period of occupation under an agreement to grant such a lease. A disregard clause may even extend to improvements carried out by a person who held as deemed lessee or as a licensee prior to taking a lease, even if there had been no actual agreement for a lease.[169] A different type of clause has required the disregard of the effect on rental values of improvements carried out by the classes of persons listed in section 34 of the Landlord and Tenant Act 1954 Part II. There have been difficulties in deciding whether a clause of this type refers to the originally-enacted version of this particular provision or to the extended version applying as from 1969.[170]

C – Method of disregarding improvements

In determining the effect on the rental value of tenants' improvements, as distinct from repairs, for the purpose of disregarding the former, it seems that a valuer is entitled to adopt any appropriate valuation method which is not erroneous in law. He may take an unimpeachable comparable and deduct the rent of

---

[166] *cont.*
Investment Co Ltd v *Havering London Borough Council* [1992] 1 EGLR 155; also *Ipswich Town Football Club Co Ltd* v *Ipswich Borough Council* [1988] 2 EGLR 146.
[167] Law Society/ISVA Model Clause Version B.
[168] If the tenant carries out improvements to which the landlord could have reasonably withheld his consent, the disregard clause will not, it seems, apply: *Hamish Cathie Travel England* v *Insight International Tours Ltd* [1986] 1 EGLR 244.
[169] Otherwise the work may be taken into account: see *Euston Centre Properties Ltd* v *H&J Wilson Ltd* [1982] 1 EGLR 57; *Historic Houses Hotels Ltd* v *Cadogan Estates* [1993] 2 EGLR 151.
[170] As to which see *Brett* v *Brett Essex Golf Club Ltd* [1986] 1 EGLR 154, CA.

the comparable from the rent of the subject premises to find out the value of the improvement. If the valuer decides to deduct the capital cost of the improvement as at the review date, a process which has not been condemned outright by the High Court, some allowances seemingly need to be made, for inflation, and to reflect the fact that, during the term of the lease, the benefit of the improvement is likely gradually to be written off.[171] In the case of major works of improvement, the landlord may reap some financial benefits in due course. There would seem to be no good reason, therefore, why he should be able to charge the tenant at rent review for that work in an increased rent.

---

[171] *GREA Real Property Investments Ltd* v *Williams* [1979] 1 EGLR 121; *Estates Projects Ltd* v *Greenwich London Borough Council* [1979] 2 EGLR 85.

Chapter 6

Landlords' remedies for tenants' breaches of repairing obligation

I – Introduction

We here review the scope of the landlord's remedies for tenants' breaches of repairing obligation and some problems connected with them. Where a tenant breaks his repairing obligations, the landlord may claim damages. Owing to the effect of statute,[1] the landlord cannot recover damages to any greater extent than the loss or "diminution" he proves to his reversion. This provision has given rise to complications, as where there is an actual or even an assumed market for the premises even though they may be badly out of repair. Thus, in a recent case,[2] the landlords had let a former textile mill, part of which was a listed building, on a series of leases to tenants who left the premises in so poor a state of repair that remedial work was agreed to cost some £312,500. The High Court refused to award this sum and decided that there would be a notional buyer for the premises – a person who would hold the property for a short period while deciding whether to refurbish or redevelop it. The state of disrepair would be a factor that person would take into account, as affecting short lettings. The sum of £40,000 was awarded. The landlord had not proved that the damage to his reversion exceeded this sum. At the same time, because the breach was taken to have some effect on the hypothetical potential buyer, its effect could not be ignored completely.

The landlord may also be subject to the limitations of the Leasehold Property (Repairs) Act 1938 if he carries out tenants' repairs, having entered the premises under an express right to do so, and then claims the cost from the tenant, where the lease does not expressly allow him to take this latter step. The courts have now ruled that if the landlord is entitled under the terms of the

---

[1] Landlord and Tenant Act 1927, s 18(1).
[2] *Craven (Builders) Ltd* v *Secretary of State for Health* [2000] 1 EGLR 128.

lease not only to enter the premises and to do repairs, but also to recover the cost of the work, the 1938 Act does not apply.³ In this way, circumvention of the 1938 Act is rendered a relatively easy matter: equally, it is hard to see why recovery of costs clauses should come within its mischief. A decision of the High Court allowing landlords in what at present appear to be exceptional cases, to claim specific performance of the tenant's repairing obligations⁴ has also increased the range of landlord remedies.

II – Impact of damages of Leasehold Property (Repairs) Act 1938

If a landlord claims damages for breach of a tenant's covenant to keep or put in repair, which is commenced at any time when three years or more of the lease remain unexpired, he cannot do so unless he serves on the tenant not less than one month before commencing the action a notice under section 146(1) of the Law of Property Act 1925.⁵

A – Scope of expression "damages"

This particular statutory rule as to damages claims, which operates as an absolute bar if it is not complied with, was enforced against an unfortunate landlord who carried out repairing work to the outside of the premises concerned. However, the lease, while allowing entry onto the premises, did not in terms allow the landlord to recover from the tenant the cost of the remedial work. Since no statutory warning notice had been served prior to the execution of the work on the tenant, the landlord's claim to recover the cost of the work as damages failed.⁶ It was not apparent that there was any oppressive conduct by the landlords within the mischief of the 1938 Act, which was aimed at preventing oppressive forfeiture notices served by speculators buying up leasehold property in poor repair and then enforcing heavy dilapidations claims which tenants cannot meet.⁷

---

3   *Jervis* v *Harris* [1996] 1 EGLR 78, CA.
4   *Rainbow Estates Ltd* v *Tokenhold Ltd* [1998] 2 EGLR 34.
5   Leasehold Property (Repairs) Act 1938, s 1(2). The tenant then has a right to serve a counter-notice on the landlord, as described in Chapter 7 of this book.
6   *SEDAC Investments Ltd* v *Tanner* [1982] 3 All ER 646.
7   *National Real Estate & Finance Co Ltd* v *Hassan* [1939] 2 KB 61 at 78, CA.

B – Default covenants

Where there is a state of disrepair in breach of covenant by the tenant, but the lease contains an express recovery of costs clause or default covenant, the 1938 Act notice requirement has no application. If a tenant fails to execute repairs for which he is liable under the lease, the landlord may have an express right, following non-compliance with a landlord's notice to the tenant to repair, to enter the premises, carry out the work and then to charge the tenant with its cost. The liability of a tenant in default in these circumstances is classified as being a contract debt, not damages. The landlord is said to be claiming a sum spent in remedying the tenant's breach, to avoid his reversion suffering lasting loss.[8] The Court of Appeal made these rulings in a case where the landlord was entitled to serve on a 999-year lessee a three-month notice to repair. On the latter's failure to comply with the notice, the landlord was expressly entitled to enter the premises, carry out works and recover the cost on demand from the tenant. The Court of Appeal rejected the approach of an earlier High Court decision,[9] in which a recovery of costs clause had been stigmatised as a device designed to remove from the tenant the opportunity given to her by the lease to carry out repairs with contractors of her choice. The Court of Appeal held that a recovery of costs clause was outside the mischief of the 1938 Act. Such clauses had been in existence before the Act was passed. If it had been intended to provide against them, very different language was to be expected from the legislature than it in fact used. Indeed, "it was not the intention of Parliament to put obstacles in the way of a landlord whose object is to secure that necessary repairs are carried out, preferably at the expense of the tenant, but if necessary his own".[10]

However useful these clauses may be to landlords, they might work oppressively.[11] In the absence of direct authority, it seems that

---

[8] *Jervis v Harris* [1996] 1 EGLR 78, CA.
[9] *Swallow Securities Ltd v Brand* [1981] 2 EGLR 48.
[10] Per Millett LJ in *Jervis v Harris* [1996] 1 EGLR 78 at 81H.
[11] If a landlord indulged in expenditure on work which was useless to the tenant and of minimal value to him, this would appear to fall outside a recovery of costs clause, and by inference within the mischief of the 1938 Act, as it is not expenditure which the landlord genuinely intended should be incurred: per Millett LJ in *Jervis v Harris, supra* at 81F–G.

a landlord may only recover for works to a reasonable standard, as he has control over the methods of work and the contractors who will execute it.[12] Landlords' entry and repair clauses are likely to be construed narrowly rather than widely.[13] A landlord making use of a recovery of costs clause would presumably also waive any right to forfeit the lease.[14] If a landlord carried out works beyond the scope of the tenant's covenant, or not as clearly specified in his notice, he could not recover the cost to that extent under a recovery of costs clause. The fact that the landlord has first to spend his own money on the work before trying to recover it from the tenant was, however, taken as some disincentive to misuse by him of his rights under an entry and repair clause.[15]

C – Statutory limits on liability

Safeguards are conferred for former tenants and their guarantors, against whom the landlord may wish to claim under a recovery of costs clause, as a result of the section 17 of the Landlord and Tenant (Covenants) Act 1995, which applies both to "old" and "new" tenancies – those granted both prior to or on or after January 1 1996. Section 17(6)(c) of the 1995 Act refers to "any amount payable under a tenant covenant of the tenancy providing for the payment of a liquidated sum in the event of a failure to comply with any such covenant". A default covenant would seem to fall within this definition, whether it is widely construed so as to include any tenant covenant in the tenancy or is narrowly construed so as only to apply to a secondary obligation to pay a liquidated sum on default.[16]

The landlord is thus in principle only able to obtain a refund under a recovery of costs clause, not recovered from the current

---

[12] *Plough Investments Ltd* v *Manchester City Council* [1989] 1 EGLR 244 at 247–248.
[13] See *Amsprop Trading Ltd* v *Harris Distribution Ltd* [1997] 2 EGLR 78 at 82L.
[14] Mitchell and Williams (1996) 140 Sol J 768.
[15] *Hamilton* v *Martell Securities Ltd* [1984] Ch 266 at 279; approved in *Jervis* v *Harris* [1996] 1 EGLR 78 at 81E.
[16] Fancourt, para 20–08, favours the wider of the two constructions of s 17(6)(c), but appears to think that if the narrower view is adopted, the tenant's obligation to refund the landlord should appear as part of a single clause in his covenants, otherwise the obligation to repay might lie outside s 17(6)(c).

tenant, against a former tenant or his guarantor if he complies strictly with the requirements of section 17. In particular, he must serve a prescribed form notice[17] on the former tenant, or, if he seeks recovery from the latter, on any guarantor of his,[18] specifying details of the claim. The notice must be served within a six-month period of the debt becoming due – which, depending on the way the lease is drafted, could be the date the landlord spends the money, or the date of his demand for payment, or the date of service of his notice on the tenant.[19] In the absence of authority, it is thought that the six-month time limit would be strictly construed against the landlord.

III – Section 18(1) of the Landlord and Tenant Act 1927

A – Introduction

At common law, where the landlord brought an action for damages at the end of the term for breaches of a lessee's covenant to repair, the lessee had to pay for the reasonable and proper cost of putting the premises into repair so as literally to comply with the repairing covenant, as originally contemplated at the date of the lease.[20] It was thought necessary expressly to restrict the landlord to any actual loss he had suffered,[21] and so section 18(1) of the Landlord and Tenant Act 1927 was passed. The courts have construed this provision so that, as far as possible, it does not deprive a landlord of compensation for any actual losses which he proves he has suffered by the lessee's breach.[22] Indeed, it is open to question whether the common law and the 1927 Act pose different principles in assessing landlords' damages for breach of covenant to repair.[23] The aim of both is to compensate the landlord for his losses

---

[17] Form 1 of the Landlord and Tenant (Covenants) Act 1995 (Notices) Regulations 1995, SI 1995 No 2964.
[18] *Cheverall Estates Ltd* v *Harris* [1998] 1 EGLR 27.
[19] Fancourt, para 20.12.
[20] *Joyner* v *Weeks* [1891] 2 QB 31, CA.
[21] As noted by the Law Commission, Law Com No 238 (1996), para 9.37.
[22] See *Ultraworth Ltd* v *General Accident Fire and Life* [2000] 2 EGLR 115 (no loss there proved).
[23] See Law Com No 238, *supra*, para 9.37; Hill and Redman, para 1187; also *James* v *Hutton and J Cook & Sons Ltd* [1950] 1 KB 9 (damages for breach of lessee's covenant to restore premises based on ordinary compensatory principles).

occasioned by the tenant's breach. However, the Law Commission decided, despite the fact that section 18(1) had probably been overtaken by subsequent developments in the general common law, not to recommend any changes to it.[24]

B – Section 18(1) of Landlord and Tenant Act 1927

The first limb of section 18(1) of the 1927 Act provides that:

> Damages for a breach of covenant or agreement to keep or put premises in repair during the currency of a lease, or to leave or put premises in repair at the termination of a lease, whether such covenant or agreement is express or implied, and whether general or specific, shall in no case exceed the amount (if any) by which the value of the reversion (whether immediate or not) in the premises is diminished owing to the breach of such covenant or agreement...

This provision imposes a ceiling of loss or diminution to the landlord's reversion[25] on the maximum amount of damages recoverable.[26] It makes the task of establishing loss to the landlord's reversion a matter of some difficulty where the cost of remedial repairs is not the appropriate measure of loss, as where the landlord has not carried out remedial work and intends to sell or has sold the premises at the expiry of the lease. In some cases, the landlord is able to base his claim on the cost of repairs, adding a sum for loss of rent while the work is under way,[27] the court reducing the claim to the fall in the value of the reversion if this is less than the cost of repairs.[28] The onus of proving injury to his

---

[24] Report, *supra*, para 9.39. In Scotland, there is no statutory ceiling on damages corresponding to s 18(1) of the 1927 Act (see eg Gordon, 19–209ff; McAllister, p 34). This might call into question the retention of any ceiling on damages in England and Wales.

[25] ie, the immediate reversion: *Terroni* v *Corsini* [1931] 1 Ch 515.

[26] As recognised by Luxmoore J in *Hanson* v *Newman* [1934] Ch 298. No allowance could be given to the tenant on account of the fact that, as there happened, the reversion was accelerated by forfeiture.

[27] In *Drummond* v *S&U Stores Ltd* [1981] 1 EGLR 42, the landlords were allowed to add to their claim the cost of VAT on repairs which they, not being VAT registered, could not recover; see also *Elite Investments Ltd* v *TI Bainbridge Silencers Ltd (No 2)* [1987] 2 EGLR 50; but as noted by Woodfall, 13.082, VAT paid could only now be claimed as damages by residential landlords, since business landlords can charge VAT on rent.

[28] As in *Craven (Builders) Ltd* v *Secretary of State for Health* [2000] 1 EGLR 128.

reversion is on the landlord, who must therefore adduce proper evidence, especially in relation to claims arising against the tenant near the beginning or in the middle of the term of the lease, otherwise the landlord runs the risk that the court will conclude that there was no serious damage to his reversion. If the tenant defends a damages claim at the end of the lease, and fails to produce evidence that the diminution to the reversion is much less than the cost of repairs, he runs a "serious risk" that the court will accept that cost, or that cost as slightly discounted, as the best evidence of diminution.[29]

C – Damages recoverable during term against lessee

If the landlord claims damages during the term and does not base his claim on the cost of repairs, the statutory ceiling may restrict the amount of damages recoverable by the landlord. It does not follow that because no repairs are going to be done by the landlord that the diminution in value of the reversion must be only nominal.[30] The maximum sum recoverable may be the loss of capital market value to the reversion caused by the breach, which will presumably vary in proportion to the length of the residue of the term at the date of the landlord's claim.[31] As the contractual expiry date of the lease approaches, the cost of repairs necessary to relet the premises may be pleaded as evidence of the injury to the value of the landlord's reversion.

In the case of a breach of covenant by a building lessee to construct new buildings on a vacant site by a certain date, which has passed, the amount of damages for breach recoverable by the landlord is the difference between the value to him of the freehold of the undeveloped site as at the date of the breach and the (notional) value of the site with buildings put up at that date. However, in one case,[32] certain deductions from the basic amount

---

[29] *Crewe Services & Investment Corporation v Silk* [1998] 2 EGLR 1 at 5B, CA. On the facts, concerned with a dilapidated farm, the case was described as a "difficult intermediate case".
[30] *Culworth Estates Ltd v Society of Licensed Victuallers* [1991] 2 EGLR 54 at 56D (Dillon LJ).
[31] *Doe d Worcester School Trustees v Rowlands* (1841) 9 C&P 734; also *Smiley v Townshend* [1950] 2 KB 311, CA.
[32] *Lansdowne Rodway Estates Ltd v Potown Ltd* [1984] 2 EGLR 80.

were made, so as not to award the landlord the benefit of any anticipated remuneration from the notional buildings which accrues to the lessee. The court, in arriving at the final award of damages, took into account: (1) the design and quality of the buildings; (2) their location; (3) the security of the capital and income; and (4) the pattern of income and the prospects of an increase. Since it now appears that a breach of covenant to build by a certain date is not necessarily incapable of remedy, and may be complied with late, it could be argued that, if the lessee does eventually comply with his covenant, the landlord's damages might be reduced by section 18(1) to a purely nominal sum.

D – Damages recoverable after the lease expires

Where damages are claimed by the landlord after expiry of the lease, the impact of section 18(1) of the 1927 Act varies, depending on whether the landlord intends to relet the premises after he has paid for the cost of remedial work or whether he intends to dispose of them as by sale.

Repairs to re-let premises

If the landlord carries out reasonably necessary repairs to the premises with a view to reletting them for the same purpose as under the previous lease, the cost of such repairs is good evidence of the damage to the landlord's reversion within section 18(1).[33] In a leading case,[34] the landlord let certain rooms in a house on a one-year residential tenancy. The tenant had covenanted to deliver up the premises in good tenantable repair. The tenant quit, leaving the interior in a bad state of repair. The landlord obtained an award of £36 in damages based on the cost to him of redecorating so as to be able to relet. The Court of Appeal held that in a simple case where necessary repairs to enable reletting were carried out, the cost of these should be the starting-point of any award. There was no finding of the capital value of the house and of the part let to the tenant but such evidence was not required as the rooms would not

---

[33] See eg *Shortlands Investments Ltd* v *Cargill plc* [1995] 1 EGLR 51, where the court had regard to the costings in a landlords' schedule of dilapidations.
[34] *Jones* v *Herxheimer* [1950] 2 KB 106, CA.

be sold off apart from the house itself. In a more recent case,[35] an interesting aspect was that the damage to the reversion was found to have exceeded the cost of repairs. The premises had been sold after lease end for £320,000; the damages were computed at £134,083. This sum, the evidence of diminution to the landlord's reversion, was awarded to them. The fact that the landlord was not going to do repairs did not entail that his damages had to be nominal.

If a landlord lets dilapidated premises to a new lessee who undertakes to carry out repairs, in return for reimbursement by him, the landlord may still claim damages on account of dilapidations from the erstwhile tenant.[36] If a landlord pays new lessees compensation for their carrying out repairs, as in the form of a reverse premium, the fact of payment may be evidence of his loss on account of the state of the premises, subject to questions of the effect of the statutory ceiling. Thus, the fact that the landlords of dilapidated premises had paid new incoming tenants £690,000 as an estimate of the cost of bringing the premises up to condition was evidence of damage to the landlord's reversion. However, because the actual dilapidations were greater than the disrepair for which the outgoing tenant was responsible, the only way to assess the part of the difference for which the tenants were liable was to assess the cost of repairs and then to apply the statutory "cap" to it.[37]

This principle would not apply if a new lessee intends to carry out work which would render any repairs redundant, as shown by the result in a High Court case. The landlord granted a long lease of dilapidated premises to a new tenant who covenanted, in return for an initial rent reduction, to carry out substantial improvements, costing twice the value of any repairs. He failed to prove that the value of his reversion had been diminished by the erstwhile lessee's breach, depriving him of a dilapidations claim of £271,211.[38]

---

[35] *Culworth Estates Ltd* v *Society of Licensed Victuallers* [1991] 2 EGLR 54, CA. The premises had a value for sale as a single unit.
[36] *Haviland* v *Long* [1952] 2 QB 80, CA.
[37] *Shortlands Investments Ltd* v *Cargill plc* [1995] 1 EGLR 51. The damage to the reversion was assessed as the difference between the amount a willing transferor would have paid to a willing transferee of the premises in a proper state of repair and the amount which would be paid if the premises were handed over in their dilapidated state.
[38] *Mather* v *Barclays Bank plc* [1987] 2 EGLR 254.

Comparable values and scarcity

Sometimes, to assess the loss to the landlord's reversion, if this is less than the cost of repairs, relevant comparable values may be taken. However, judicial notice has been taken that evidence of comparable commercial properties might relate to premises in repair, so that evidence of the value of comparables out of repair might be very difficult indeed to obtain.[39] If there is no relevant comparable evidence, the basis of assessment may, as seen, be taken as the cost of putting the property into repair as set out in a schedule of dilapidations.

The erstwhile lessee cannot use a scarcity factor as a way of reducing the damages otherwise payable for breach of covenant. Where the premises, though out of repair, were relet to a new tenant at the same rent as the last tenant paid, it was held that the fact that there was a scarcity in demand for the relevant house in the area was an extraneous factor: damages would be assessed, subject to section 18(1) of the 1927 Act, on ordinary principles. The landlord was not limited to the modest sum he spent to make the premises presentable.[40] At the same time, as appears, if a landlord sells premises and the price paid is unaffected by the state of repair of the premises, there is likely to be no damage to the landlord's reversion and he recovers no damages. These matters indicate that there is little difference in the operation in ordinary cases as between statute and the common law: both apply compensatory principles to landlord's damages claims.

Special factors

Section 18(1) may, however, deprive landlords of damages where premises are occupied by business tenants protected by Part II of the Landlord and Tenant Act 1954 and entitled to new tenancies. Section 34 of this Act postulates that the landlord is entitled to the open market rent of the premises. This hypothesis, applying for statutory renewal purposes, has the side-effect that the landlord's reversion is deemed, no matter that the premises may be in fact be out of repair, to suffer no loss.[41] As a result, a landlord was

---

[39] *Drummond* v *S&U Stores Ltd* [1981] 1 EGLR 42 at 44D.
[40] *Jaquin* v *Holland* [1960] 1 All ER 402, CA.
[41] *Family Management* v *Gray* [1980] 1 EGLR 46, CA.

permitted only to lead evidence of damage to his reversion based on any non-correspondence of the old repairing obligations in the lease and subleases granted prior to statutory renewal and in relation to the non-occupied parts of the premises.[42]

A further possibility of no deemed loss to the reversion arises where the premises cannot, after the lease has expired, be relet for the same purposes as they were let under the previous lease. For example, the premises might originally have been let for residential purposes, with a high standard of repair and decoration, but at the date of the action they cannot, because of planning restrictions or for some other reason, be relet for those purposes but for commercial purposes, with a different standard of repairs.

In one case,[43] a house was let on a 19-year lease and at the end of the lease it was left out of repair by the tenant. The house could not be relet for residential purposes. Damages were awarded to the landlord but these were less than the amount he claimed as the cost of repairs necessary literally to comply with the covenant to repair. There was no evidence of damage to the reversion: the property was therefore valued in notional repair and in its condition out of repair and the landlord was awarded the difference between the two sums.

Capital value unaffected

If the landlord can show that the price paid or expected to be paid for his reversion on sale in the open market is affected by the state of repair of the premises, the abatement in price, actual or notional, would on the face of it constitute the diminution in the value of his reversion.[44] The statutory ceiling may eliminate the landlord's claim to damages if he has sold the premises at a market price, the purchaser being unaffected by the state of repair of the premises.[45] This happened with certain premises which sold for a "good price" for conversion into two flats and two maisonettes. The landlord failed to show that the value of his reversion had been diminished

---

[42] *Crown Estate Commissioners* v *Town Investments Ltd* [1992] 1 EGLR 61.
[43] *Portman* v *Latta* [1942] WN 97.
[44] *Re King* [1962] 1 WLR 632 at 647 (Buckley J).
[45] *London County Freehold and Leasehold Properties Ltd* v *Wallis-Whiddett* [1950] WN 180 (compulsory purchase of premises, taking premises off the market, and no proof of drop in price on account of disrepair); also *Ultraworth Ltd* v *General Accident Fire and Life* [2000] 2 EGLR 115.

by the breach of covenant of the erstwhile lessee.[46] If the development value of the premises is shown to exceed any rental or investment value they may have, the landlord might be unable to show loss to the value of his reversion: however, the High Court awarded a landlord the difference between the site value and the investment and development value of the subject premises.[47] If no buyer can be found on the relevant date, so that there is no "market" in that sense for the premises, the court may have to decide itself what a willing buyer would be prepared to pay for them, even if there is no evidence as to the cost of repairs.[48]

E – Covenants outside section 18(1)

Section 18(1) of the 1927 Act does not apply to a covenant either to spend a stated sum on repairs and decorations or to pay the landlord the difference betwen the stated sum and the amount actually expended, because such amounts are not classified as damages.[49] Nor does section 18(1) affect the amount of damages recoverable where there is a breach of covenant by the tenant not to alter the demised premises. The measure of damages is at large but will not necessarily be the cost of reinstatement of the premises. Thus, where an assignee of a lease under a covenant not to alter the internal planning of the premises sublet parts of the premises and the subtenants converted the premises into five separate flats, the landlord was entitled to the cost of reinstatement as damages, although the premises as flats were more valuable than they had been as a single dwelling-house.[50]

---

[46] *Landeau* v *Marchbank* [1949] 2 All ER 172. The landlord recovered nominal damages only.
[47] *Shane* v *Runwell Ltd* [1967] EGCD 88.
[48] *Craven (Builders) Ltd* v *Secretary of State for Health* [2000] 1 EGLR 128 (£40,000 awarded on basis that only hypothetical buyer would hold premises for an interim period, so as to let on short tenancies, attributing some value to repairs: a schedule of dilapidations was not even *prima facie* evidence of diminution, since the landlord did not intend to carry out repairs).
[49] *Moss' Empires Ltd* v *Olympia (Liverpool) Ltd* [1939] AC 544.
[50] *Eyre* v *Rea* [1947] KB 567; also *James* v *Hutton and J Cook & Sons Ltd* [1950] 1 KB 9 (where nominal damages were awarded, loss not having been proved to the letting value).

F – Special rules

In a number of circumstances the application of section 18(1) of the 1927 Act has given rise to particular problems.

Requisitioned premises

Where the demised premises are requisitioned, the relevant date for the assessment of damages[51] is the date when the lease in fact terminates, even though the premises, at that date, remain requisitioned with the result that the landlord is not in a position to resume possession. The measure of damages is the difference in the value of the reversion at the termination of the lease, between the premises in their then state of disrepair and in the state which they would have been in had the covenants been fulfilled. If the requisitioning authority has made alterations to the premises prior to the termination of the lease which render any repairs covenanted for valueless, then the tenant is not liable in damages to that extent; if the authority prior to the date of termination of the lease makes good any disrepair, the tenant is entitled to the benefit of that work in reduction of damages payable,[52] no doubt in accordance with the compensatory principle.

Protected tenancies

A special rule, of diminishing significance with the advent as from January 15 1989 of assured tenancies, applies where the premises are held by a protected (or statutory) tenant under the Rent Act 1977. Where a house was occupied by a protected tenant and was badly out of repair, the measure of damages was not the agreed cost of putting the house into the state of repair required by the lease, which was originally granted for a long term, but based on comparing the difference in price if the house were sold, subject to the protected tenancy, both in and out of repair.[53]

---

[51] In relation only to tenants' dilapidations during the period when the premises are not requisitioned: Landlord and Tenant (Requisitioned Land) Act 1944, s 1.
[52] *Smiley v Townshend* [1950] 2 KB 311, CA.
[53] *Jeffs v West London Property Corporation* [1954] CLY 1807.

Subleases

Difficulties may arise where the whole or part of the premises are sublet. If the sublease expires soon before the expiry date of the head lease, the cost of putting the property into repair at the expiry of the head lease may be the appropriate measure of diminution of value of the freeholder's reversion for statutory purposes.[54] However, if the sublessee's obligation to repair is less onerous than that of his own landlord, or if the commencement date of his term is later than that of the head lessee's own lease, the quantum of damages the latter may recover from the sublessee may not necessarily be the same as that recoverable by the freeholder from the head lessee. If, however, the sublessee has caused the dilapidations, his immediate landlord may be able to claim the full amount of his own loss as paid to the freeholder under an indemnity clause from the sublessee, if an indemnity clause is indeed inserted in the sublease.[55]

If the sublessee has broken his obligation to repair, and the freeholder claims damages from the head lessee, who holds only a nominal reversion, the value of the latter would seem to be nil or even negative. Hence, he cannot literally prove any diminution to the value of his reversion as against the sublessee.[56] The courts have avoided this result by ruling that the diminution in value is the minus value of the reversion, caused by the dilapidations in question, so that the 1927 Act does not of itself debar the intermediate landlord's claim against his sublessee.[57] Proof of a diminution in this sense is required: a mesne landlord with a 15-day nominal reversion failed to make out his claim where it appeared that the sublessee had also bought out the freehold of the premises in question.[58]

Impossibility of performance

In view of the recent development of a doctrine of frustratory mitigation, a tenant who is unable, due to planning or wartime

---

[54] *Ebbetts* v *Conquest* [1895] 2 Ch 377, CA.
[55] No such right is implied: see *Wholesale Invisible Mending Co* v *Needle* [1971] CLY 6557.
[56] As noted by Hill and Redman, para 6689.
[57] *Lloyds Bank Ltd* v *Lake* [1961] 1 WLR 884.
[58] *Espir* v *Basil Street Hotel Ltd* [1936] 3 All ER 91, CA.

restrictions, to comply with a covenant to rebuild could now claim, at least if the restrictions were in force at the date of the landlord's claim, that no damages should be awarded.[59]

G – Demolition or alteration of demised premises

Until the passing of the 1927 Act, any damages payable by a tenant for breach of his covenant to repair at the end of the term were unaffected by the fact that the buildings were to be demolished as soon as the lease ended.[60] Moreover, such damages were payable even if the repairs would be nullified by structural alterations which were in contemplation at the end of the lease.[61] The severity of these rules explains the second limb of section 18(1) of the Landlord and Tenant Act 1927 which provides:

> ... no damage shall be recovered for a breach of any such covenant to leave or put premises in repair at the termination of a lease, if it is shown that the premises, in whatever state of repair they might be, would at or shortly after the termination of the tenancy have been or be pulled down, or such structural alterations made therein as would render valueless the repairs covered by the covenant or agreement.

The tenant has a complete defence, the onus of proving which, in contrast to the first part of section 18(1), is on him,[62] if he proves that the landlord intends completely to demolish the premises or completely to nullify the value of any repairs by structural alterations. If all the landlord intends to do is to carry out more modest structural alterations which reduce the value of any necessary repairs, the damages payable by the tenant are reduced.[63]

Relevant date

The relevant date for intended demolition or structural alterations is the date of the termination of the lease,[64] and any subsequent

---

[59] *John Lewis Properties plc* v *Viscount Chelsea* [1993] 2 EGLR 77.
[60] *Rawlings* v *Morgan* (1865) 18 CB (NS) 776.
[61] *Inderwick* v *Leach* (1885) 1 TLR 484.
[62] *Crown Estate Commissioners* v *Town Investments Ltd* [1992] 1 EGLR 61 at 64M–65A.
[63] *Fairclough (TM) & Sons* v *Berliner* [1931] 1 Ch 60.
[64] Whether by expiry, repossession following forfeiture, physical re-entry of the premises or reletting to a different tenant.

change of mind by the landlord is not relevant in assessing his intention.[65] Put in a different way, in assessing the requisite intention, events subsequent to the termination of the lease are not relevant, and the court will assume that the intention of the landlord as communicated to the tenant as at lease end is the sole relevant intention.[66]

The landlord may have plans for demolition or structural alterations at the end of the lease. He must be shown definitely to have made up his mind at the relevant date: if there is a sufficiently formidable succession of fences to be surmounted before his intention can be realised then the landlord does not "intend" the project and the tenant has no defence based on the second limb of section 18(1) of the 1927 Act. Where a landlord had a project for redevelopment of the site in question but, at the date of termination of the lease, had to obtain planning permission and a building licence and to determine the financial viability of the project, the tenant could not escape liability in damages under section 18(1) as the landlord's project was too provisional.[67] The court does not expect to go into all the minute details of the implementation of the landlord's project. If the landlord's plans are reasonably firm and he has a reasonable prospect of implementing these, with any necessary planning permissions and no significant obstacles posed to him by the imposition of any planning conditions, the tenant should be able to establish the necessary "intention" of the landlord.[68]

IV – Specific performance against the tenant

It was decided in an old case[69] that a landlord could not obtain a decree of specific performance against a tenant in order to force

---

[65] *Salisbury* v *Gilmore* [1942] 2 KB 38, CA.
[66] *Keats* v *Graham* [1959] 3 All ER 919, CA.
[67] *Cunliffe* v *Goodman* [1950] 2 KB 237, CA.
[68] A local authority tenant was unable to use compulsory purchase powers over unfit premises to avoid paying a landlord damages, where these had been proved, because section 18(1) of the 1927 Act could not be used to allow a tenant to profit from its own wrong, in having committed a breach of repairing covenant: *Hibernian Property Co Ltd* v *Liverpool Corporation* [1973] 2 All ER 1117.
[69] *Hill* v *Barclay* (1810) 16 Ves 402.

him to carry out specified repairs which he has failed to execute in breach of his covenant. However, at that time the tenant could not obtain the remedy against his landlord. This is no longer the case. Moreover, despite recent dicta doubting the availability of this remedy to a landlord,[70] this ancient ruling was stigmatised as being in effect redundant.[71]

The High Court has now ruled that it has a jurisdiction in equity to order specific performance to a landlord against a defaulting tenant.[72] There were no constraints of principle or authority against granting the remedy to the lessor. The court would need to be able to define clearly the work to be done, and could examine the end result of the works, both factors overcoming any problems arising out of the need for constant supervision, which was once thought to debar the remedy. There was held to be a need for great caution in granting the remedy to the landlord. It would be a rare case when the remedy would be appropriate, since the primary remedies of a commercial landlord remained forfeiture or entry and repair. There was an overriding need to prevent injustice or oppression to the tenant, so that the remedy could not be used to circumvent the Leasehold Property (Repairs) Act 1938, even though the 1938 Act does not apply to the remedy of specific performance. In considering this issue, it would seem that the court might pay regard to the sort of factors set out in section 1(5) of that Act – all of which would appear to preclude the routine use of specific performance as a means of enforcing tenant repairing obligations.

In the case under consideration, the court decreed specific performance because the lease contained no re-entry proviso, so that it could not be forfeited; there was no right in the landlord to enter and repair. There was a serious state of disrepair of the deteriorating property in the order of £300,000; and the defendants' means were unknown or slight. Local authority repair notices had been served in relation to the premises. The works were such that they could with certainty be specified in the court's order. Even if section 1 of the 1938 Act had applied by analogy, it was held that several of its leave grounds could have been made out on the facts.

---

[70] *Regional Properties Ltd* v *City of London Real Property Co Ltd* [1981] 1 EGLR 33 at 34 (Oliver J).
[71] Law Com No 238 (1996), para 9.20.
[72] *Rainbow Estates Ltd* v *Tokenhold Ltd* [1998] 2 EGLR 34; see HW Wilkinson (1998) NLJ, October 9 1998; also Pawlowski and Brown [1998] Conv 495.

The facts of this case seem exceptional. A lease without a re-entry clause is unusual – this one fact, combined with the gravity of the breaches, seemed to justify invoking the remedy of specific performance. At the same time, it is thought unlikely that landlords will succeed in obtaining this relief as an alternative remedy to damages, or on the same more ready basis as do residential tenants holding short leases. If the landlord has a loss which can be quantified, then on ordinary equity principles, damages would seem adequate in all save exceptional cases.

Chapter 7

Forfeiture for breach of covenant to repair

I – General principles

A – Introduction

Preliminary considerations

The landlord is entitled to forfeit a lease for breaches by the tenant of covenant to repair, provided the lease expressly reserves a right of re-entry or forfeiture for any breach of covenant. If a lease is forfeited and no relief granted to the tenant or other applicant such as a sublessee or mortgagee, the landlord regains possession of the premises.

This Chapter examines the rules, once described as being a "legal minefield",[1] which govern forfeiture by the landlord for breaches of tenants' covenants to repair. The law leans against forfeiture. The remedy is supposed to be of last, not first resort – to be successful only in the case of serious and unremedied breaches of covenant to repair by an occupying tenant, or where the tenant has left the premises, and there is no prospect of the breaches being cured by someone else such as a legal mortgagee. A number of principles combine to make forfeiture difficult. To anticipate, the common-law doctrine of waiver, which preceded the restrictions in the Law of Property Act 1925 and the Leasehold Property (Repairs) Act 1938, prevents a landlord from forfeiting a lease for breach of covenant to repair where he has demanded or accepted rent for a post-breach period with the requisite knowledge of the breach. If a landlord wishes to forfeit the lease, he is bound to serve on the tenant a notice complying with statute (Law of Property Act 1925 s 146). He cannot then enforce his claim to forfeiture unless he allows the tenant a reasonable time in which to remedy the breach. In addition, special rules applying to breaches of covenant to repair require that the landlord proves that the tenant knew of the service of this notice (as where it was sent in the registered post or by

---

[1] *Rexhaven Ltd* v *Nurse and Alliance and Leicester Building Society* (1995) 28 HLR 241 at 255.

recorded delivery). The tenant may also stay the forfeiture proceedings by resort to the Leasehold Property (Repairs) Act 1938, whose general result is to compel the landlord to prove a sufficiently serious case to be allowed to regain possession of the premises. Even then, the court has power within the 1938 Act to allow the tenant further time to comply with his covenant. In any event, the tenant can apply to the court, under its statutory jurisdiction, for relief against forfeiture (Law of Property Act 1925 s 146(2)). There are no fixed rules as to the granting of relief, but it is ordinarily granted if the tenant is willing and financially able to remedy established breaches of covenant within a reasonably short time. The grant of relief restores the tenant to the same position as if no forfeiture had been incurred. Unless the lease expressly entitles the landlord to recover costs incurred during a claim for forfeiture, statute limits his rights, notably to a case where the tenant obtains relief against forfeiture.

At the same time, where the premises are commercial, and vacant, the landlord may, instead of using proceedings, peaceably re-enter, as by reletting the premises.[2] The lease will then be forfeited, unless the re-entry is vitiated, as where there is a residential occupier present, or because no suficient time had previously been allowed to the tenant to remedy the breaches, and subject to issues of relief. Peaceable re-entry has sometimes been used by commercial landlords, no doubt owing to its cheapness and speed, when compared to taking proceedings. Its use is no panacea, as where there is a mortgagee, who is in a position to execute the repairs concerned and who has applied to court for relief against forfeiture notwithstanding the landlord's peaceable re-entry.

Reform

Some time ago, the Law Commission proposed reforms to forfeiture. Under these, one of whose many advantages would be to provide a uniform set of rules in this field, a termination order scheme would be enacted.[3] The doctrine of waiver would be abolished and replaced by a principle that if the landlord caused a reasonable tenant to believe that he would not ask for a termination order, he would be unable to seek one. If the court decided,

---

[2] As in *Re AGB Research plc* [1994] EGCS 73.
[3] Law Com No 142 (1985).

following proceedings, to terminate a lease for breach of covenant to repair, it could, in a serious case, at least where there was no prospect of remedy by the tenant, grant an absolute termination order. Alternatively, it could order the conditional termination of the lease, which the tenant could avert by remedial action.[4] In 1994, the Commission, with a view to prompting legislation, produced a Termination of Tenancies Bill.[5] In the case of proceedings for breach of "obligation" to repair, clause 12 of the Bill would, following the original reform proposals, preserve a special notice procedure. It requires service of a landlords' preliminary notice on the tenant and entitles the tenant to serve a counter-notice. A leave requirement and leave grounds would be also retained from the present law by clause 12. The government have recently indicated that implementation of the 1994 reform package is unlikely, but it is considering further restrictions on commencing forfeiture proceedings in the long residential sector.[6] In essence, the proposals, which would be brought in when a suitable opportunity arose, would involve the separation of proceedings to determine facts and those involving forfeiture. The former proceedings would go before a leasehold valuation tribunal and the latter to the courts. The proposals would build on existing legislation.[7] Thus in relation to alleged breaches of covenants to repair or not to carry out structural alterations, the landlord would have first to serve a notice on the leaseholder in which 30 days to respond were given. Details of the breach would have to be set out. If the breach was remedied or reasonable compensation paid to the landlord within the 30 days, the landlord would not be able to take any further action, and the lessee would not be liable to any costs of the landlord in serving the notice, overriding any contrary provision in the lease. Any dispute would be referred to a leasehold valuation tribunal.[8] The government thought that this suggested reform

---

[4] See Cherryman (1987) 84 LS Gaz 1042; also PF Smith [1986] Conv 165. There is summary of the main proposals in Megarry and Wade para 14–167, commenting that "this scheme would be a very considerable improvement on the present law and would remove most of the difficulties to which it has given rise". It is hard to disagree.

[5] Law Com No 221; see HW Wilkinson [1994] Conv 177.

[6] Commonhold and Leasehold Reform Draft Bill and Consultation Paper (2000) Cmnd 4843 (August 2000), Part II, Section 4.6, para 8.

[7] ie Housing Act 1996, s 81.

[8] Commonhold and Leasehold Reform Draft Bill and Consultation Paper Part II (2000) Section 4.6, paras 15 and 16.

would deter unscupulous landlords from exploiting leaseholders' fears about the costs of legal proceedings – even though they also thought that forfeiture itself rarely occurred in practice.[9]

Leave of court

A forfeiture action requires the leave of the court, if brought against a company tenant which is subject to an administration order under the Insolvency Act 1986 Part II (s 11(1)(c)). At one time, there was a conflict in the cases as to whether such leave is required where the landlord forfeits the lease by peaceably re-entering the premises.[10] The prevailing view was that peaceable re-entry is treated as a self-help remedy analagous to distress for unpaid rent. The latter, in the case of commercial leases, did not require the assistance and so not the leave of the court to render it effective;[11] on this basis, neither did peaceable re-entry.[12] However, leave of the court is now required (s11(3)(ba), added by Insolvency Act 2000, s 9(3)).

Peaceable re-entry: some additional considerations

The use by any party to a lease of a self-help remedy carries with it an inherent danger of violence. Statute[13] renders it a criminal offence to use or to threaten violence for the purpose of regaining entry to premises. However, if commercial premises are vacant, at the time of the entry, a landlord's changing the locks does not render the entry any less peaceable. Parliament, when it repealed the statutes of forcible entry, opened the way to this particular method of regaining possession. Residential tenants are protected from peaceable re-entry by section 2 of the Protection from Eviction Act 1977. In the case of premises let as a dwelling-house, it is not lawful to enforce a right of re-entry or forfeiture otherwise than by proceedings in court while any person is lawfully residing in them or in any part of them.

---

[9] *Ibid*, Section 4.6, paras 2 and 3.
[10] In *Re Olympia & York Canary Wharf Ltd* [1993] BCLC 154 it was held that no leave was required; the contrary was held to be the case in *Exchange Travel Agency* v *Triton Property Trust plc* [1991] 2 EGLR 50.
[11] *McMullen & Sons Ltd* v *Cerrone* [1994] 1 EGLR 99.
[12] *In re Debtors no 13A10 and 14A10 of 1994* [1995] 2 EGLR 33 (Insolvency Act 1986, s 252(1)); also *In re Lomax Leisure Ltd* [1999] 2 EGLR 37.
[13] Criminal Law Act 1977, s 6(1).

Peaceable re-entry was stigmatised by Lord Templeman[14] as being dubious and dangerous. Certain lessors had, by means of a "dawn raid", gained entry at 6am to premises. At that time, the premises were temporarily unoccupied. In breach of covenant, the lessees were converting part of the property into short-term letting accommodation. The House of Lords held that a tenant who had been subjected to peaceable re-entry was entitled to apply to the court for relief under statute without any formal time-limit. By contrast, if a landlord lawfully regains actual possession after having proceeded against the tenant by writ, his re-possession is a complete bar to the tenant applying for relief under statute.

However, business landlords still appear to be making use of peaceable re-entry.[15] The remedy is said to be cheap when compared to proceedings. The Law Commission has recently recommended that peaceable re-entry for vacant premises should be incorporated into new legislation, once the law applying to forfeiture is reformed, albeit in a somewhat circumscribed form.[16] That would preserve the ability of commercial landlords to resort to self-help in the forfeiture field but at least some of the uncertainties of the law might be clarified.

B – Waiver of breach

Even before the statutory restrictions on forfeiture were enacted, the common law had a crude instrument for restricting forfeitures. This is the doctrine of waiver. Under it, the landlord cannot forfeit the lease if he has expressly or impliedly waived a breach of covenant to repair. Essentially there are three component elements to waiver, although, as noted by the Court of Appeal, "this is not an area of law where any rigid or precise taxonomy of principles is possible".[17]

---

[14] In *Billson v Residential Apartments Ltd* [1992] 1 EGLR 43 at 44L (as compared to the "civilised" method of using a writ).
[15] According to Megarry and Wade, para 14–123.
[16] Consultative Document, "Termination of Tenancies by Physical Re-Entry", Law Commission (1998); press release June 30 1999. As to whether forfeiture by peaceable re-entry would contravene the Human Rights Act 1998, see Bruce "Barring Peaceable Re-Entry" 150 (2000) NLJ 462.
[17] *Ballard (Kent) Ltd v Oliver Ashworth (Holdings) Ltd* [1999] 2 EGLR 23 at 27A (Robert Walker LJ).

1. The alleged act of waiver must unequivocally recognise the subsistence of the lease.
2. The landlord must know of the breach of covenant from which the right of re-entry arises at the time of the alleged act of waiver.
3. The act of recognition of the lease must be communicated to the tenant.[18]

Silence of the landlord is not a waiver. Waiver is judged objectively. The landlord's actions count for more than his words or disclaimers. Thus, if he has demanded rent knowing of the tenant's breach of repairing covenants, the landlord cannot avoid his having affirmed the lease even if he had expressly reserved his rights.[19]

The landlord may expressly waive a breach by a statement to the tenant to the effect that he will not bring forfeiture proceedings. Nor may he act in an inconsistent manner, as where a landlord demanded rent due after the relevant breach had come to his knowledge, which is commonly relied on as implied waiver, but he claimed to act "without prejudice" to his right to forfeit the lease.[20] The effect of waiver on covenants to repair is, however, mitigated owing to the fact that in contrast to a breach of a negative covenant such as not to alter the demised premises, a breach of covenant to repair is of a continuing nature.[21] If the tenant continues in breach by the time the next rent payment is due, the landlord has a fresh cause of action against him, with a fresh right to elect for forfeiture or to affirm the lease by waiver.[22] A forfeiture action may be based on the whole of the events leading down to the current state of affairs, even if the landlord had, at an earlier date, waived his right to forfeit. In the case of a breach of an absolute covenant against structural alterations, however, if the landlord waives the breach, he puts it out of his power to forfeit the lease on account of that

---

[18] *Cornillie* v *Saha* (1996) 72 P&CR 147, CA.
[19] *Ballard (Kent)* v *Oliver Ashworth (Holdings) Ltd*, *supra*.
[20] *Segal Securities Ltd* v *Thoseby* [1963] 1 QB 887.
[21] As shown by the fact that where a landlord gave the tenant notice to remedy a breach within three months, he did not waive the breach of covenant by demanding rent three days after the notice ran out: *Penton* v *Barnett* [1898] 1 QB 276; *Greenwich London Borough Council* v *Discreet Selling Estates Ltd* [1990] 2 EGLR 65, CA.
[22] *Expert Clothing Service & Sales Ltd* v *Hillgate House Ltd* [1985] 2 EGLR 85.

breach, which is classified as a once and for all breach whose consequences cannot be undone.

The degree of knowledge required as a precondition of waiver is not of all the details of the breach. A landlord who, though denied access to the premises, had discovered from photographs and other observations that works in breach of covenant (against alterations) were well under way was put on his election.[23]

The conduct lessees may seek to rely on as amounting to implied waiver is a rent demand or acceptance of rent for a period due after the breach is known to have taken place. Proof of either action by the landlord, or of a distress for rent due after the breach, will deprive him of the right to forfeit, at least for the time being. As a landlord who receives regular rent payments into a bank account might thereby be in danger of implied waiver, provided that he returns the payment promptly to the lessee, once informed by his banker of the fact of the payment, the landlord may be able to avert an automatic waiver.[24]

II – Status of lessee during forfeiture proceedings

Once a claim form (formerly known as a writ) unequivocally claiming forfeiture and possession has been served, if an unconditional possession order is made, the forfeiture is backdated to the date of the claim for possession.[25] Once the landlord claims to forfeit a lease, no waiver of any breaches of the tenant's repairing covenant is possible, owing to the fact that the landlord is taken to have elected unequivocally to forfeit the lease. If possession is unconditionally ordered, the tenant covenants in the lease or any sublease are extinguished.[26] During the "limbo period" from issue of the claim form until such time as possession is ordered, tenant repairing covenants in the lease remain in some sense alive, at least vis-à-vis sublessees. A lessee facing a forfeiture claim was ordered, by mandatory injunction, to comply with his covenant with sublessees

---

[23] *Iperion Investments Corporation* v *Broadwalk House Residents Ltd* [1992] 2 EGLR 235.
[24] *John Lewis Properties plc* v *Viscount Chelsea* [1993] 2 EGLR 77.
[25] *Ivory Gate Ltd* v *Spetale* [1998] 2 EGLR 43, CA; *Maryland Estates Ltd* v *Bar Joseph* [1998] 2 EGLR 47, CA.
[26] *Ivory Gate Ltd* v *Spetale, supra; Twogates Properties Ltd* v *Birmingham Midshires Building Society* [1997] EGCS 55, CA (subleases).

to put a lift in working order.[27] At the same time, it appears that during the period between the issue of a writ (or claim form) seeking forfeiture of a lease and an order for possession (or the grant of relief against forfeiture to the tenant as the case may be) the landlord cannot enforce repairing covenants against the tenant.[28]

The complexities of this part of the law are not eased by the fact that both where there is a genuine claim to relief against forfeiture, and during the period leading down to the making of an order for possession or the dismissal of the claim or the grant of relief against forfeiture, the lease is said to have a "trance-like" existence. "None can assert with assurance whether it is alive or dead".[29] It has "somewhat obscure" characteristics.[30] The tenant retains an interest in the premises until the landlord regains physical possession under a court order, or the lease is reinstated following the grant of relief against forfeiture, or the action fails for some other reason or is withdrawn.[31] However, once the landlord lawfully regains actual possession of the demised premises, the tenant cannot any longer apply for relief under statute against forfeiture. There is no inherent jurisdiction to grant him relief outside statute, which is taken to provide a comprehensive code.[32]

III – Statutory notice requirements and relief against forfeiture

A – Requirement of written statutory notice

Preliminary

The tenant, assuming he is still occupying the premises is entitled, is the person best able to decide how to deal with the alleged

---

[27] *Peninsular Maritime Ltd* v *Padseal Ltd* [1981] 2 EGLR 43, CA.
[28] *Associated Deliveries Ltd* v *Harrison* [1984] 2 EGLR 76, CA. In such a case, the landlord would presumably have to pray in aid the law of waste if the tenant damaged the premises during the "limbo period". Nor can the tenant enforce repairing covenants during the "twilight period" against the landlord: *Lambeth London Borough Council* v *Rogers* [2000] 1 EGLR 28, CA.
[29] *Meadows* v *Clerical, Medical and General Life Assurance Society* [1981] Ch 70 at p 75B (Megarry V–C).
[30] *Liverpool Properties Ltd* v *Oldbridge Investments Ltd* [1985] 2 EGLR 111 at p 112H (Parker LJ).
[31] *Hynes* v *Twinsectra Ltd* [1995] 2 EGLR 69, CA.
[32] *Smith* v *Metropolitan City Properties Ltd* [1986] 1 EGLR 52.

breaches of repairing covenant, to have a warning of the landlord's intention to forfeit the lease, and to receive particulars of the allegations which the landlord makes against him. To that end, a notice complying with section 146(1) of the Law of Property Act 1925 must be served on the lessee by the landlord prior to issue of any claim form claiming possession and forfeiture or to any peaceable re-entry. The notice must be in writing (s 196(1) of the 1925 Act). It may be addressed to the lessee by designation (s 196(2)) – so preventing unnecessary inquiries as to the lessee's correct or latest name. The notice may be left at the lessee's last-known place of abode or business,[33] or affixed, or left for him, on the land or any house or building comprised in the land.

Proof of service on lessee

Thanks to section 18(2) of the Landlord and Tenant Act 1927, the landlord must prove that the fact of service of a section 146 notice was known, in particular, to the lessee. Otherwise the landlord cannot forfeit the lease by proceedings or by peaceable re-entry. This requirement is specific to alleged breaches of covenants to keep or put premises in repair or to leave or put them in repair at the end of the lease. However, a statutory notice may be sent by registered post or recorded delivery – in which case the landlord benefits from a statutory presumption of due delivery to the tenant. This deems the lessee, unless he proves the contrary, to have knowledge of the fact of service as from the time the notice would have been delivered in the ordinary course of the post.[34]

Contents of section 146(1) notice

The contents of a statutory forfeiture notice are as follows.

(a) It must specify the breach complained of.
(b) If the breach is capable of remedy, the notice must require the lessee to remedy the breach.

---

[33] As opposed to a place where the lessee never resided nor was an occupier: *Willowgreen Ltd v Smithers* [1994] 1 EGLR 107, CA.
[34] See *Lex Service plc v Johns* [1990] 1 EGLR 92 at 95A–B (Glidewell LJ) – non-receipt of the notice insufficient: proof is required eg that the notice was returned to the landlord, or that it was not acknowledged as received by the recipient.

(c) It must also require the lessee to make compensation for the breach.

Only if the lessee fails, within a reasonable time of the notice, to remedy the breach, may the landlord proceed with forfeiture. Once he receives a section 146(1) notice, the lessee is able to consider his position. He may decide to comply with the notice and to carry out any uncontested repairs specified in any landlords' schedule of dilapidations. He is also entitled to apply for relief against forfeiture, which he may do as soon as he receives a statutory notice,[35] since it is at this stage that the landlord is, within section 146(1) of the 1925 Act, "proceeding" to enforce a right of re-entry or forfeiture. Where a landlord opts for peaceable re-entry of commercial premises this right of the lessee could be of significance to him.

A section 146(1) notice will be invalid unless it clearly informs the lessee of his right, by counter-notice, to claim the Leasehold Property (Repairs) Act 1938. This aspect is further considered below: suffice it to note here that, if the 1938 Act is claimed within 28 days of service of the main notice, the forfeiture proceedings are stayed. These can only be revived by the landlord proving a serious case within the confines of the 1938 Act, justifying forfeiture.

If the lease is held by an assignee in possession, he, and not the original tenant, is entitled to be served with the relevant section 146(1) notice, and to claim the 1938 Act, even if the assignment was contrary to the lease. It is the assignee, because he is in possession, who is in a position to decide what, if anything, to do about the notice.[36] For the same reason, where a lessee had left the premises and his mortgagee was in possession (having obtained an order for possession for mortgage arrears against the lessee), the mortgagee was entitled to service of a landlord's section 146(1) notice and to claim the benefit of the 1938 Act.[37] A section 146(1) notice must, where appropriate, be served on all joint lessees.[38]

Although section 146(1) of the 1925 Act enables a landlord to claim compensation, he is entitled to omit any such claim from a

---

[35] *Pakwood Transport* v *15 Beauchamp Place* (1977) 36 P&CR 112, CA.
[36] *Old Grovebury Manor Farm Ltd* v *W Seymour Plant Sales & Hire Ltd (No 2)* [1979] 3 All ER 504.
[37] *Target Home Loans Ltd* v *Iza Ltd* [2000] 1 EGLR 23.
[38] *Blewett* v *Blewett* [1936] 2 All ER 188.

statutory notice.[39] If relief against forfeiture is granted to the lessee, it seems therefore that the landlord cannot claim compensation not sought in his notice. Should the landlord realise that his notice is invalid, he is entitled, as soon as he realises the true position, to serve a second notice on the tenant, and need not await the determination of the validity of his first notice.[40]

Specification of breaches

A section 146(1) notice should set out the covenants to repair. It should then set out the specific breaches of covenant which will be relied on as grounds for forfeiture. These may appear in a Schedule of Dilapidations appended to the main notice. A landlord was held entitled to state that his schedule was interim, reserving his rights in respect of any other defects which might exist.[41] No reasonable objection could be taken to this course of action, since it was on the specified breaches on which the landlord had based his forfeiture claims.

Under section 146(1) of the 1925 Act, the notice must require that the alleged breaches are to be remedied. Hence, a notice which asserted that the tenant had broken his covenants for repairing the inside and outside of houses was bad, since it failed to specify any particular alleged breaches.[42] By contrast, a schedule which indicated general repairs to be done in all of six houses covered by a forfeiture notice as well as specified repairs in some of the houses, all works listed under general headings, and without the landlord having specified the precise manner in which repairs must be carried out, was upheld.[43] Lord Buckmaster said this.[44]

> I can find nowhere in the section any words which cast upon the landlord the obligation of telling the tenant what it is that he must do. All that the landlord is bound to do is to state particulars of the breach of covenant of which he complains and call upon the lessee to remedy

---

[39] *Rugby School (Governors)* v *Tannahill* [1935] 1 KB 87, CA.
[40] *Fuller* v *Judy Properties Ltd* [1992] 1 EGLR 75, CA.
[41] *Greenwich London Borough Council* v *Discreet Selling Estates Ltd* [1990] 2 EGLR 65, CA.
[42] *Fletcher* v *Nokes* [1897] 1 Ch 271.
[43] *Fox* v *Jolly* [1916] 1 AC 1, HL.
[44] In *Fox* v *Jolly*, *supra* at 11, applied as the "crucial passage" in *Adagio Properties Ltd* v *Ansari* [1998] 2 EGLR 69 at 71K.

them. The means by which the breach is to be remedied is a matter for the lessee and not for the lessor. In many cases specification of the breach will of itself suggest the only possible remedy.

The statutory notice and any accompanying schedule is being given to a person who knows or ought to know of the condition of the premises.[45] At all events, in connection with a section 146(1) notice, which was upheld against tenants who were alleged to have broken a covenant against alterations by converting one flat into two studio flats, the Court of Appeal insisted that a notice, provided it told the tenant what he had done to convert the premises and what he must do to reinstate the premises to one flat, did not have to be overburdened with details of the particular items of work alleged to have broken the covenant.[46]

Technical aspects of notices

Some care must be taken in the drawing up of statutory notices, since a notice which set out a painting covenant not in fact in the lease was held bad.[47] A notice which alleged several breaches of covenant, of which one related to the unhygienic condition of a toilet, could have been servered and forfeiture proceedings based on the particular breaches alleged of repairing covenants.[48] A tenant's claim that a notice was invalid merely because it related both to alleged breaches of the covenant to repair and to repaint, on the sole ground that he had disproved a breach of the covenant to paint failed: the rest of the notice was held good.[49]

Reasonable time to remedy

The landlord is not required to state in his notice a time within which the alleged breach or breaches complained of must be remedied. However, a forfeiture may only be proceeded with if, following a section 146(1) notice, the lessee fails after a reasonable time to carry out uncontested repairs as specified in the notice. He

---

[45] Per Lord Atkinson in *Fox* v *Jolly, supra* at 17.
[46] *Adagio Properties Ltd* v *Ansari, supra*.
[47] *Guillemard* v *Silverthorne* (1908) 99 LT 584.
[48] *Starrokate Ltd* v *Burry* [1983] 1 EGLR 56 at p 57H (May LJ); also *Silvester* v *Ostrowska* [1959] 1 WLR 1060.
[49] *Pannell* v *City of London Brewery Co* [1900] 1 Ch 496.

may avert forfeiture by complying with the requirements of the notice within the time allowed by the notice or otherwise – irrespective of his past record of compliance. The landlord must estimate the amount of the reasonable time, and if he makes an error, and issues a claim form or re-enters peaceably too soon, the forfeiture process is invalid. In one case, only two months elapsed between service of the relevant notice and the landlord's peaceable re-entry, which was held to be far too short a time period on the facts.[50] The period of reasonable time is not, however, governed by any rule of law. A reasonable time must, however, be allowed for a remedy of all alleged breaches.[51] By way of further example, it was held that a minimum period of at least three months should have been allowed by the landlord to elapse, as from service of his forfeiture notice, where the dilapidations schedule there particularised the carrying out of major works.[52]

B – Relief against forfeiture

Introduction

The tenant may apply for relief against forfeiture under statute either in the landlord's action, or in a separate action of his own. If the landlord proceeds by claim form, the tenant cannot apply for relief once the landlord has lawfully regained possession of the premises, but the lessee may ask for relief, without limit of time, should the landlord decide peaceably to re-enter the premises.[53] The effect of relief, if granted, is to reinstate the tenant as though no forfeiture had taken place.[54]

Where the tenant has claimed the Leasehold Property (Repairs) Act 1938, the landlord must pursue his claim within the confines of section 1(5) of the Act and prove his case within one of the narrow statutory leave gateways. It is open to the court itself to specify the breaches to be remedied, or, under section 1(6) of the Act, to

---

[50] *Target Home Loans Ltd v Iza Ltd* [2000] 1 EGLR 23 (where it seems that access to the premises was denied to the mortgagee's surveyor).
[51] *Hopley v Tarvin Parish Council* (1910) 74 JP 209.
[52] *Bhojwani v Kingsley Investment Trust Ltd* [1992] 2 EGLR 70.
[53] *Billson v Residential Apartments Ltd* [1992] 1 EGLR 43, HL.
[54] *Hynes v Twinsectra Ltd* [1995] 2 EGLR 69, CA (where this result was said to follow without any formal grant of relief, where forfeiture proceedings had been compromised).

adjourn or dismiss the application on condition that certain repairs are done. Nothing, in this, precludes a relief application by the tenant under section 146(2) of the 1925 Act.[55]

Terms of statutory jurisdiction to grant relief

The statutory jurisdiction to grant relief, which cannot be contracted out of, is conferred by section 146(2) of the Law of Property Act 1925. This provides as follows:

> Where a lessor is proceeding, by action or otherwise, to enforce such a right of re-entry or forfeiture, the lessee may ... apply to the court for relief; and the court may grant or refuse relief, as the court, having regard to the proceedings and conduct of the parties ... and to all the other circumstances, thinks fit; and in case of relief may grant it on such terms, if any, as to costs, expenses, damages, compensation, penalty, or otherwise, including the granting of an injunction to restrain any like breach in the future, as the court, in the circumstances of each case, thinks fit.

Principles on which relief granted

There are no fixed rules as to the circumstances in which, in a given case of breach of covenant to repair, the court will grant relief in its discretion.[56] Equity leans against forfeiture. It is not now correct to say that relief will only be granted in an exceptional case.[57] Provided the landlord suffers no significant or irremediable loss or damage to his reversion and is adequately compensated for any breaches, and is not going to be saddled with a patently undesirable tenant, the tenant has in principle every prospect of obtaining relief. The court may impose terms on the tenant as a condition of the grant of relief, to that extent protecting the landlord's legitimate interests. The facts and circumstances of each case are considered on their merits, including the conduct of both parties. It may be that a landlord who resorts to the uncivilised remedy of peaceable re-entry would find it harder to resist a relief

---

[55] *Associated British Ports* v *CH Bailey plc* [1990] 1 EGLR 77, HL.
[56] *Hyman* v *Rose* [1912] AC 623 at 631 (Earl Loreburn LC).
[57] *Southern Depot Co Ltd* v *British Railways Board* [1990] 2 EGLR 39 at 44A; *Crown Estate Commissioners* v *Signet Group plc* [1996] 2 EGLR 200 at 209K.; *Mount Cook Land Ltd* v *Hartley* [2000] EGCS 26.

application than a lessor who proceeds in court.[58] Equally, a tenant who shows every wish not to comply with a covenant which he has wilfully broken stands little chance of obtaining relief,[59] as would a lessee who refused without good cause to comply with any conditions imposed on the granting of relief.

Ordinarily, therefore, a lessee who is willing and financially able to remedy breaches of covenant to repair which have been proved by the landlord,[60] should avert forfeiture. The court may well lay down a time-limit for compliance with the lessee's obligations as a condition of relief, which may be extended once or even more frequently at discretion on the tenant's application.[61] The lessee will ordinarily have to pay the landlord compensation for unremedied breaches. If, after expiry of the time specified in the section 146(1) notice, but by the date of the hearing, the tenant has complied with his covenant to repair, he should obtain relief. Exceptional circumstances may, however, render such a course of action inappropriate, as where the breaches have been gross or wilful, or where it appears that the tenant is not likely to comply with his obligations in the future.[62]

Because forfeiting a lease for a substantial term may inflict a penalty out of proportion to the gravity of the tenant's breach of covenant, even if serious, the value of the tenant's interest may be a relevant factor. If the tenant can remedy the breaches at a cost which is relatively small compared to the value of his interest, he strengthens the case for relief, especially where there is no lasting

---

[58] *Billson* v *Residential Apartments Ltd* [1993] EGCS 150.
[59] *Darlington Borough Council* v *Denmark Chemists Ltd* [1993] 1 EGLR 62, CA.
[60] In *Target Home Loans Ltd* v *Iza Ltd* [2000] 1 EGLR 23, some repairs had been carried out by a new short-term tenant of the premises, which to that extent made inaccurate the landlord's schedule. The court would have refused to impose compliance with the works there specified as a condition of relief against forfeiture.
[61] The Court of Appeal is reluctant to interfere with the proper exercise of a county court discretion in relation to granting additional extensions of time, as shown in *Crawford* v *Clarke* [2000] EGCS 33, CA, where a judge had refused a third time extension to a lessee, influenced by the latter's "prolonged and lamentable" failure to fulfil the relief terms, even though his decision was described as "unusual".
[62] *Cremin* v *Barjack Properties Ltd* [1985] 1 EGLR 30 at 32, CA.

damage to the landlord's reversion.[63] If the state of disrepair of the premises, or a wilful failure to comply with a covenant to build by a certain date, cause permanent and irretrievable damage to the landlord's reversion, that is a good ground for refusing relief.[64] But if no lasting damage is caused to the landlord by the breach, as where a lessee could not build by a certain date owing to unforeseen difficulties, a grant of relief would, even in that case, depending on all the circumstances, not seem inappropriate.[65]

A lessee or sublessee who deliberately breaks an absolute covenant against the making of structural alterations to the premises may find it hard to persuade the court to grant him relief, as where a sublessee in deliberate breach was refused relief but his innocent mesne landlord was granted it, on terms that the premises should, within a specified but extensible time-limit, be reinstated.[66] The fact that a lessee commits a deliberate breach of this covenant is not a complete bar to relief if the circumstances otherwise favour granting it, as where the premises may be readily reinstated, or where the head landlord could not have objected to the sublessee's alterations to his part of the premises, on which a large sum had been spent.[67] A lessee who undertook major structural alterations to parts of the premises, in flagrant breach of covenant and in contravention of planning restrictions, which seriously damaged the building, was refused relief as there was no suggestion that he would comply with any conditions of the grant of relief.[68]

C – Relief to underlessees and mortgagees

Notification by landlord

If a head lease is forfeited, any sublease granted out of it and any mortgage is destroyed, as from the court's order. Any sublessee

---

[63] See *Ropemaker Properties Ltd* v *Noonhaven Ltd* [1989] 2 EGLR 50 (where there were breaches of negative covenant. Relief was granted to a lessee, who until the breaches took place had been an excellent tenant, owing in part to the high value of his long lease in a prime site).
[64] *Southern Depot Co Ltd* v *British Railways Board* [1990] 2 EGLR 39.
[65] *Underground (Civil Engineering) Ltd* v *London Borough of Croydon* [1990] EGCS 40.
[66] *Duke of Westminster* v *Swinton* [1948] 1 KB 524.
[67] See *Iperion Investments Corporation* v *Broadwalk House Residents Ltd* [1992] 2 EGLR 235.
[68] *Billson* v *Residential Apartments Ltd* [1993] EGCS 150.

then becomes a trespasser with no right to exclude the landlord from the premises.[69] Because of these drastic consequences for derivative interest holders, in forfeiture proceedings, where the claimant landlord knows[70] of any person entitled to claim relief against forfeiture as underlessee (which includes as mortgagee) under section 146(4) of the Law of Property Act 1925, the claim particulars must give the name and address of that person. The landlord must file a copy of these particulars for service on the mortgagee or underlessee concerned.[71]

These procedural rules have no application where the landlord has physically re-entered the premises without, on purpose, informing any mortgagee or sublessee. In such a case, the first intimation a sublessee or mortgagee has of forfeiture proceedings might be on his being informed by the Land Registry that his title has been deleted from the relevant register. The landlord is not "proceeding" within section 146(4) of the 1925 Act, since the forfeiture has taken place. The holder of a derivative interest may be able to invoke a residual equity jurisdiction to grant relief on terms. The High Court acceded to such an application by a mortgagee faced with the loss of a security worth some £17,600.[72] Though this was a case relating to arrears of rent and service charges, the enactment of section 146(4), which, in contrast to the rest of section 146, applies to non-payment of rent as well as to breaches of repairing obligations, is arguably not exclusive in the area to which it applies of a wider equity jurisdiction outside the scope of the 1925 Act. Exceptional cases apart, the existence of a statutory jurisdiction to grant relief precludes the co-existence of any general inherent equity jurisdiction. Where a mortgagee or sublessee is duly served with a copy of the writ (or new claim form) concerned, he must apply within the limits of the statutory relieving jurisdiction, or lose any right to any relief.[73]

---

[69] *Viscount Chelsea* v *Hutchinson* [1994] 2 EGLR 48, CA.
[70] As where he has been notified as required by the terms of the lease by his lessee.
[71] Civil Procedure Rules 1998 16PD-002.
[72] *Abbey National Building Society* v *Maybeech Ltd* [1985] Ch 190. The existence of any or any general jurisdiction was denied *obiter* in *Billson* v *Residential Apartments Ltd* [1992] 1 EGLR 43 at 45, HL and also in the Court of Appeal [1991] 1 EGLR 70 at 76. See Bridge [1992] CLJ 216; also PF Smith [1992] Conv 33.
[73] *United Dominions Trust Ltd* v *Shellpoint Trustees Ltd* [1993] 4 All ER 310, CA.

Statutory power to grant relief

The court is given power by section 146(4) of the Law of Property Act 1925 to grant relief to a derivative interest holder in the form of an order vesting the residue of the lease in the applicant. The grant of relief is a matter for the discretion of the court, but it is ordinarily subject to conditions which may deter sublessee applicants (at least those with short terms) from seeking relief in view of the expense of compliance when compared to the length of the term vested in them. It is ordinarily a condition precedent of the grant of a vesting order that the applicant pays off all outstanding rent arrears and remedies, within a specified timetable, all breaches of covenant to repair. This would seem to be the case even if the disrepair affects a greater part of the premises than that subdemised or mortgaged,[74] since the governing principle is to restore the landlord to the same position that he would have been in if no breaches of repairing covenant had taken place. The applicant may be required to pay off landlords' costs not recovered from the lessee.

In addition, it is generally a condition of a vesting order that the applicant must undertake the same, or at least as stringent, repairing obligations as those in the erstwhile head lease.[75] Thus, relief was refused to the subtenant of a basement, part of the premises demised to a head tenant whose lease was forfeited, because the applicant refused to pay for repairs, which were required to put the basement into good repair, at a cost of £10,000 to £15,000. The former sublessee was not prepared to undertake the same onerous repairing obligations as those in the head lease, under a mere monthly tenancy, which was all that he was entitled to obtain.[76]

Nature of relief

Relief takes the form of a new lease which is vested by the court's order, taking effect as if a lease had been executed by the legal estate owner, in the applicant.[77] The term vested cannot exceed the

---

[74] *Chatham Empire Theatre (1955) Ltd* v *Ultrans Ltd* [1961] 1 WLR 817 (rent arrears).
[75] *Creery* v *Summersell* [1949] Ch 751 at 767.
[76] *Hill* v *Griffin* [1987] 1 EGLR 81, CA.
[77] See *Official Custodian for Charities* v *Mackey* [1985] Ch 168 at 185G.

duration of the original sublease. The new lease takes effect as from the date of the court's order. In the case of the jurisdiction to grant relief applying to covenants to repair,[78] it is not retrospective. If the landlord had at some earlier time re-entered, and at a later date a vesting order is made, it will have no effect on acts done by the landlord between his actual re-entry and the date of the order. Thus a mortgagee who obtains a vesting order cannot demand that the landlord pays to him mesne profits received by him as from the date he forfeited the lease down to the date of vesting.[79]

However, a mortgagee by subdemise and an underlessee are also now entitled to make application for relief against forfeiture, at least in the case of a lease which has not yet been terminated, under section 146(2) of the 1925 Act, owing to the wide definition in section 146(5)(b) of the expression "lessee".[80] There may be advantages in opting for an application under section 146(2), even though the conditions for the obtaining of relief would not materially differ from those applying to applications under section 146(4), so as to prevent the landlord suffering any unfair prejudice. Since the effect of the grant of relief against forfeiture under section 146(2) is, by contrast to relief under section 146(4), retrospectively to re-instate the lease, as though no forfeiture had occurred, the applicant has no obligation to pay mesne profits to the landlord, but only rent, which might be a lower sum if the market value of the premises at the date of the granting of relief is greater than the passing rent.

D – Recovery of landlords' costs

Certain landlords' costs incurred in forfeiture procedings are recoverable. By section 146(3) of the Law of Property Act 1925:

> A lessor shall be entitled to recover as a debt due to him from a lessee, and in addition to damages (if any), all reasonable costs and expenses

---

[78] As opposed to breaches of covenant to pay rent, where the effect of relief under s 146(4) of the 1925 Act is retrospective: *Bank of Ireland Home Mortgages* v *South Lodge Developments* [1996] 1 EGLR 91 at 93B.
[79] See *Official Custodian for Charities* v *Mackey (No 2)* [1985] 1 EGLR 46, CA; *Pellicano* v *MEPC plc* [1994] 1 EGLR 104.
[80] *Escalus Properties Ltd* v *Robinson* [1995] 2 EGLR 23, CA. A mortgagee by legal charge would seem to be entitled to opt for a relief application under s 146(2) of the 1925 Act, owing to the words "deriving title under a lessee" in s 146(5)(b).

properly incurred by the lessor in the employment of a solicitor and surveyor or valuer, or otherwise, in reference to any breach giving rise to a right of re-entry or forfeiture which, at the request of the lessee, is waived by the lessor, or from which the lessee is relieved, under the provisions of this Act.

There are limits to the scope of this provision. It applies only if the tenant obtains relief against forfeiture, or where he persuades the landlord to waive the breach in question. It does not to apply where the tenant, sublessee or mortgagee remedies the breach by complying with the statutory forfeiture notice.[81] If a forfeiture takes place and the landlord in fact re-enters, section 146(3) will not apply. One way around these limits is to insert in the lease an express covenant by the tenant to pay all expenses, including solicitors' costs and surveyors' or valuers' fees, incurred by the landlord, incidental to the preparation and service of a section 146 notice, notwithstanding that forfeiture is avoided otherwise than by relief granted by the court.[82]

If the tenant claims the benefit of the Leasehold Property (Repairs) Act 1938, the landlord's claim to costs may only be exercised if he obtains leave to proceed (s 2). This provision is limited to damages claims under section 146(3) of the 1925 Act itself. If the lease contains an express covenant by the tenant to pay solicitors', surveyors' or valuers' costs, the claim will be classed as a claim for a simple contract debt and the 1938 Act will be excluded.[83]

E – Special relief for decorative repairs

The tenant may have covenanted expressly to carry out both external and internal decorative repairs at regular intervals. A breach of covenant in relation to external decorative repairs may be

---

[81] *Nind v Nineteenth Century Building Society* [1894] 2 QB 226, CA: this ruling is not seemingly affected by the decision in *Escalus Properties Ltd v Robinson, supra*.
[82] In *Pertemps Group Ltd v Crosher & James* [1999] CLY 3676, the words "incidental to" in an express clause were held capable of extending to the fees of a building surveyor employed to cost proposed works after the service of a s 146(1) notice, as well as to landlords' surveyors and solicitors fees incurred prior to the service of the notice in question.
[83] *Middlegate Properties Ltd v Gidlow-Jackson* (1977) 34 P&CR 4, CA.

a serious matter, as injury to the structure and exterior of the premises can rapidly result from non-observance of this covenant. Rigidly to insist, on pain of forfeiture, that the tenant comply with his covenant as regards interior decoration might be unreasonable where no serious physical damage to the premises could be shown from the breach. The exceptions to the provision suggest that it is aimed mainly at discouraging landlords from bringing forfeiture claims against lessees who have failed to carry out interior decorations at regular specified intervals.

By section 147(1) of the Law of Property Act 1925, which cannot be contracted out of in the lease (s 147(4)), after a notice is served on a lessee relating to internal decorative repairs to a house or other building, the lessee may apply to the court for relief and if, having regard to all the circumstances (including in particular the length of the lessee's term or interest remaining unexpired), the court is satisfied that the notice is unreasonable, it may by order wholly or partly relieve the lessee from liability for such repairs (including it seems not only liability for forfeiture but even any liability for damages). The tenant may, on receipt of the notice, immediately apply to the court for relief. Section 147(1) extends in appropriate cases to underlessees. The section is not designed to enable the tenant to break certain specific types of obligation. Hence, it is excluded in four cases (s 147(2)):

(i) Where the liability arises under an express covenant or agreement to put the property in a decorative state of repair and the covenant or agreement has never been performed.
(ii) In relation to any matter necessary or proper –

    (a) for putting or keeping the property in a sanitary condition, or
    (b) for the maintenance or preservation of the structure.

(iii) In the case of any statutory liability to keep a house in all respects reasonably fit for human habitation.
(iv) Where there is a covenant or stipulation to yield up the house or other building in a specified state of repair at the end of the term.

F – Leasehold Property (Repairs) Act 1938

The effect of the Leasehold Property (Repairs) Act 1938 upon a landlord's forfeiture action is profound. Where the 1938 Act is claimed by the tenant, all further proceedings are stayed. The threat of forfeiture is lifted, unless the landlord is able to obtain the leave of the court to proceed with his claim. The 1938 Act is an important tenants' defence against forfeiture for breaches of covenant to repair. It probably makes it not worth a landlord attempting forfeiture for breach of covenant to repair during the term of a lease, save where no other course of action is possible, even in the case of serious breaches.

Application of 1938 Act

The 1938 Act applies to residential and commercial leases, irrespective of rateable value.[84] The lease must be for a term (when originally granted) of seven years or more. At the date of the service of the landlord's forfeiture notice, three years or more of the term must remain unexpired (s 1(1) and 7(1)). The 1938 Act extends to head leases and subleases, unless their initial duration is for a term of less than seven years (s 7(1)).

The 1938 Act applies to a notice served under section 146(1) of the Law of Property Act 1925 which: "relates to a breach of a covenant or agreement to keep or put in repair during the currency of the lease all or any of the property comprised in the lease" (s 1(1)). The 1938 Act does not apply where the tenant covenants merely to leave the property in repair at the end of the lease. The Act applies only to covenants to "keep or put in repair" and so not to covenants to cleanse, even if wrongly labelled as to repair.[85] Moreover, the Act does not apply to a breach of covenant or agreement by a lessee to put in repair which is to be performed upon his taking possession or within a reasonable time thereafter (s 3). This means that an obligation to put in repair premises which are dilapidated at the commencement of the lease is excluded from the 1938 Act procedures.

Object of 1938 Act

The policy of the 1938 Act is to prevent oppressive forfeitures being brought by landlord during the currency of a substantial lease for

---

[84] It does not apply to a lease of an agricultural holding within the Agricultural Holdings Act 1986 nor to a farm business tenancy within the Agricultural Tenancies Act 1995 (s 7(1) of the 1938 Act as amended).
[85] *Starrokate* v *Burry* [1983] 1 EGLR 56.

trivial breaches of a tenant's covenant to repair. The particular mischief which the 1938 Act was designed to remedy was that of speculators buying up small property in an indifferent state of repair, and then serving a schedule of dilapidations upon the tenants, which the tenants could not comply with.[86]

The 1938 Act is not supposed to inhibit forfeiture for serious breaches of a lessee's repairing covenant. Nevertheless, it provides important relief to the lessee. He may put to an end the immediate threat of forfeiture, by means of service of a counter-notice. To overcome this, the landlord must apply for leave and then prove his case, to ordinary civil standards of proof, within one or more of the prescribed grounds at the leave application, unless the parties both agree to treat the latter as a mere formality preceding a full forfeiture hearing. Even where a landlord obtains leave to proceed, so reviving his claim to forfeit the lease, the tenant is still entitled to ask the court for relief against forfeiture, and will have knowledge of the case he has to meet and of any actions he may be able to take to avert loss of the premises.

Requirements of section 146 notice

The 1938 Act imposes strict additional requirements on forfeiture notices, where it applies. A notice under section 146(1) of the Law of Property Act 1925, which relates to a breach of a covenant or agreement to keep or put in repair during the currency of the lease, all or any of the property comprised in the lease, will be invalid unless it complies with the 1938 Act. The notice must contain a statement "in characters not less conspicuous than those used in any other part of the notice" to the effect that the lessee is entitled to serve on the lessor a counter-notice claiming the benefit of the 1938 Act. There must appear a further statement "in the like characters, specifying the time within which, and the manner in which under the Act, a counter-notice may be served and specifying the name and address for service of the lessor" (s 1(4)). Any notice or counter-notice under the 1938 Act must be in writing.[87]

These statements must be set out *verbatim*. A landlord who serves a section 146 notice, which fails to comply with the 1938 Act, may

---

[86] *National Real Estate & Finance Co Ltd v Hassan* [1939] 2 KB 61 at 78, CA.
[87] 1938 Act, s 7(2), applying s 196 of the Law of Property Act 1925, and thus incorporating its provisions as to service of notices.

later rely on the combined effect of this defective notice and a subsequent letter which fully complies with these requirements.[88] So as to discourage tenants from bringing unmeritorious technical arguments, the words "not less conspicuous" of section 1(4) have been limited to meaning "equally readable" or "equally sufficient". If a statement is no less easily readable than the rest of the notice, the whole notice will be valid even though the parts of it required to be included by the 1938 Act (say in blank spaces in a section 146 notice) are in a larger, or smaller, typeface from other parts of the notice. If a notice sets out one of a number of different methods of service of a tenant's counter-notice, it will be valid and the notice does not necesarily have to set out all the methods of service on the landlord.[89]

Counter-notice and subsequent proceedings

The 1938 Act allows a lessee who has received a notice under section 146(1) of the 1925 Act to serve, within 28 days of the service of the lessor's notice, a counter-notice on the lessor, or on the person to whom he has been paying rent, if different, to the effect that he claims the benefit of the 1938 Act (s 1(1)). The effect of a counter-notice served within the time-limits laid down by the Act[90] is that no proceedings, by action or otherwise, may be taken by the lessor for the enforcement of any right of re-entry or forfeiture under any proviso or stipulation in the lease for breach of the covenant or agreement in question otherwise than with the leave of the court (s 1(3)), generally the county court (s 6(1)). Leave may only be given if the lessor can establish one or more of five statutory grounds or "leave gateways" (s 1(5)). Otherwise the forfeiture action is put to an end by the counter-notice.

The 1938 Act had once been taken only to grant the lessee procedural rights, but the House of Lords held that, unless the parties agreed to treat the leave application as procedural only, to be followed by a full hearing, the legislation granted the lessee

---

[88] *Sidnell v Wilson* [1966] 2 QB 67, CA.
[89] *Middlegate Properties Ltd v Messimeris* [1973] 1 WLR 168, CA.
[90] By analogy with business tenancies legislation, it seems that the 28-day time-limit is strict, although in theory capable of being waived by the landlord as the party benefitted by it: *Kammins Ballrooms Co Ltd v Zenith Investments (Torquay) Ltd* [1971] AC 850.

important substantive rights.[91] As a result, it was held, the landlord must prove his full case at the leave application, and not merely a *prima facie* case which, if later proved, would justify granting leave. The landlord must therefore bring his forfeiture claim for breach of covenant to repair within one or more of the five leave "gateways" of section 1(5), which are designed to avoid forfeiture save in the case of serious breaches of covenant or other like circumstances. If the landlord obtains leave to proceed, the lessee may apply for relief against forfeiture under the 1925 Act. Alternatively, the court has power to act under section 1(6) of the 1938 Act, as where the lessee has not made an application for relief against forfeiture under the 1925 Act. The court may specify steps required to comply with the covenant, within time-limits. Equally, it has a discretion to dismiss or adjourn the leave application, provided that, where appropriate, specified repairs are carried out.

Persons entitled to serve counter-notices

The "lessee" in possession, on whom a section 146(1) notice has been served, is the person who is entitled to serve a 1938 Act counter-notice. The expression "lessee" in the 1938 Act, wherever it appears, is to be given the same meaning as in the 1925 Act.[92] Therefore, the relevant person to serve a counter-notice may be the original lessee or an assignee in possession. A legal mortgagee who had become entitled as against the lessee to possession (as holding an undischarged order for possession) was entitled to the benefit of the 1938 Act.[93] The policy of this decision is that the 1938 Act is to be available to the person who, having possession as lessee or mortgagee, is in a position to remedy any outstanding breaches. By contrast, a legal mortgagee by charge who was not in possession at the material time could not claim the benefit of the 1938 Act.[94] An assignee of the lease in possession was entitled to serve a counter-notice.[95] An assignee who, by the date of service of a section 146

---

[91] *Associated British Ports* v *CH Bailey plc* [1990] 1 EGLR 77; see Bridge [1990] CLJ 401; also PF Smith [1990] Conv 305.
[92] *Cusack-Smith* v *Gold* [1958] 2 All ER 361 at 364–365.
[93] *Target Home Loans Ltd* v *Iza Ltd* [2000] 1 EGLR 23.
[94] *Church Commissioners for England* v *Ve-Ri-Best Manufacturing Co Ltd* [1957] 1 QB 238, nor, presumably, a legal mortgagee by subdemise.
[95] *Kanda* v *Church Commissioners for England* [1958] 1 QB 332.

notice, re-assigned his lease, putting himself out of possession with no estate in the land, and so in no position to remedy any breaches, was also outside the benefit of the 1938 Act.[96]

Leave grounds

If the lessee serves a counter-notice on the landlord, the threat of a forfeiture action is lifted unless section 1(5) is satisfied. The landlord requires the leave of the court to proceed. His case must be proved to the ordinary civil standard of proof. His case must fall within specified statutory "gateways" or it fails. The landlord can rely on all or any of them.[97] The leave grounds (slightly paraphrased) are as follows.

(a) The immediate remedying of the breach is required to prevent substantial diminution in the value of the landlord's reversion, or that its value has been substantially diminished by the breach.
(b) The immediate remedying of the breach is required to give effect in relation to the premises to the purposes of any enactment, or any byelaw or other provision having effect under an enactment, or for the purpose of giving effect to any order of a court or requirement of any authority under any enactment or any such byelaw or other provision.
(c) Where the lessee is not in occupation of the whole of the premises as respects which the covenant or agreement is proposed to be enforced, the immediate remedying of the breach is required in the interests of the occupier of those premises or of part of the premises.
(d) The breach can be immediately remedied at an expense that is relatively small in comparison with the much greater expense that would probably be occasioned by postponement of the necessary work.
(e) Special circumstances exist which in the opinion of the court, render it just and equitable that leave should be given.

In a case which increased the protections of the 1938 Act for lessees, landlords sought leave to proceed under section 1(5)(a).[98] The

---

[96] *Cusack-Smith* v *Gold, supra.*
[97] *Phillips* v *Price* [1959] Ch 181.
[98] *Associated British Ports* v *CH Bailey plc* [1990] 1 EGLR 77.

premises were redundant for their original use, with apparently obsolete dock machinery, demised by a lease due to expire only in 2049. The landlords served a schedule of dilapidations on the lessees, which it would cost over £600,000 to comply with. Although the premises and their fixtures were badly out of repair, the lessees claimed that the damage to the landlord's reversion was a mere £3,500. It appeared that if the landlords could have forfeited the lease, the value of their reversion would have been enhanced. The House of Lords ruled that the landlords had, despite the fact that the premises were seriously out of repair, produced no evidence of substantial damage to their reversion in the specific factual circumstances. Their claim failed.[99] In future, a landlord must prove his case to the ordinary civil standard of proof, that is, on the balance of probabilities. Instead, these landlords seem to have assumed serious damage to their reversion from the fact of serious dilapidations.

The High Court had asserted a residual jurisdiction to refuse leave to a landlord who had proved his case under the old, low, standard of proof which no longer prevails.[100] The channelling of landlord forfeiture claims governed by the 1938 Act within the Caudine forks of the statutory leave grounds, combined with the increase in the onus of proof on landlords and the court's powers to grant relief both under the 1938 Act and under the main statutory relief jurisdiction seem to have made this particular jurisdiction obsolescent.

---

[99] Thus, compliance with the full schedule of dilapidations would serve no purpose having regard to the redundancy of the premises; the landlords would continue until lease end to collect their annual rent of £4,000, and it was difficult to accept that the landlords or any purchasers "were entitled to be frightened about what the position will be in the year 2049" – per Lord Templeman in *Associated British Ports* v *CH Bailey 0plc, supra* at 80F.

[100] See eg *Re Metropolitan Film Studio's Ltd's Application* [1962] 1 WLR 1315, also *Associated British Ports* v *CH Bailey plc* [1989] 1 EGLR 69 at 71L (at first instance).

Chapter 8

Tenants' remedies for breach of covenant to repair

In this Chapter, we examine the many remedies which a tenant has at his disposal for breach of his landlord's repairing or maintenance obligations.

I – Repudiation of lease or tenancy

At present, the tenant cannot ask the court to terminate his tenancy even if the landlord has committed serious and unremedied breaches of covenant to repair or maintain. The Law Commission recommended curing this gap in the law by means of a statutory termination order scheme, linked to a right to claim damages for loss of the tenancy.[1] On the ground of a need as then perceived, rapidly to implement forfeiture reforms, the Law Commission in 1994 produced a draft Bill which omitted to make provision for a tenants' termination order scheme.[2] As it now seems that early implementation of the 1994 proposals is unlikely,[3] it looks as if the notion of a tenants' termination order scheme is dead and buried.

However, following the example of both the High Court of Australia[4] and the Supreme Court of Canada,[5] the contractual doctrine of fundamental or repudiatory breach now appears to

---

[1] Law Com No 142, *Forfeiture of Tenancies* (1985) pp 142–157.
[2] Law Com No 221, *Termination of Tenancies Bill*; see PF Smith 'Termination of Tenancies – A Just Cause', Ch 6 in *Reform of Property Law* (1997) ed Jackson and Wilde.
[3] Commonhold and Leasehold Reform Draft Bill and Consultation Paper (August 2000) Cmnd 4843, Residential Leasehold Reform Consultation Paper, Part II, section 4.6, para 8 (hereafter "Consultation Paper (2000) Part II").
[4] *Progressive Mailing House Pty Ltd* v *Tabali Pty Ltd* (1985)157 CLR 17.
[5] *Highway Properties Ltd* v *Kelly, Douglas & Co Ltd* (1971) 17 DLR (3rd) 710.

have been accepted by the English common law.⁶ In a landmark county court decision,⁷ a three-yearly assured shorthold tenant successfully established a repudiatory breach by the landlord of his statute-implied obligations. Parts of the house concerned had become impossible to live in. The landlord had failed to carry out any remedial work despite reminders from the tenant, who accepted the fundamental breaches of the landlord by leaving the house and returning the keys.

The tenant, to establish repudiatory breach, must have given the landlord notice of the breaches, and time to remedy them. Because the landlord must be shown to have disregarded his repairing obligations to such an extent as to go to the root of the contract, the breach must be of a very grave character – such that no reasonable tenant could expect to live in the premises. It may be that a persistent refusal by the landlord to comply with his repairing obligations is the required gateway to proof by the tenant of fundamental breach. Repudiatory breach opens the way to a tenant who accepts it to claim damages from the landlord for losses flowing from the breach. In addition, the tenant ceases to be liable for future rent as from the date of the acceptance of repudiation.

II – Damages for breach of landlord's covenant to repair

A – General principles

The object of an award of damages is to put the tenant in the same position as he would have been in, if the landlord had duly performed his covenant.⁸ There is a comparison between the condition of the premises during the period of the breach and the condition they would have been in if the covenant had been performed – since damages are compensatory in nature. However, "no one inflexible form of computation [is] to be applied to all breaches".⁹

---

[6] See *Chartered Trust plc* v *Davis* [1997] 2 EGLR 83 at 88 (Henry LJ); also *Nynehead Developments Ltd* v *RH Fibreboard Containers Ltd* [1999] 1 EGLR 7.
[7] *Hussein* v *Mehlman* [1992] 2 EGLR 87; see Bright [1993] Conv 71.
[8] *Wallace* v *Manchester City Council* [1998] 3 EGLR 38, CA (where the leading cases are reviewed and which, although dealing with residential accommodation, contains "valuable statements of principle which are of general application": *Larksworth Investments* v *Temple House Ltd (No 2)* [1999] BLR 297 at 302, CA).
[9] *Credit Suisse* v *Beegas Nominees Ltd* [1994] 1 EGLR 76 at 90G (Lindsay J).

Where the landlord has broken an obligation to keep premises or specified parts of them in repair, and disrepair originates in a part of the property not demised to the tenant, he is liable in damages as from the date the disrepair occurred.[10] By contrast, where disrepair arises within the confines of the demised premises, the tenant must give notice to the landlord, who has a reasonable time from notice in which to execute repairs before liability in damages is triggered (see further Chapter 10). Where a defect is latent, this principle may deprive the tenant of any damages. The losses to the tenant must have been reasonably forseeable as the natural and probable consequence of the breach, at the time the lease was granted.[11]

The overall amount as well as the quantification of particular items of a damages award is primarily a matter for the judge hearing the case: the Court of Appeal will upset an award only if it is such as to indicate an error of principle.[12] However, where damages were assessed as a "global figure" not based on a mathematical calculation, the award was upset and a much lower figure arrived at on appeal.[13] Morritt LJ mentioned, without any need on the facts to decide the issue, since the award was well within these limits, a claim in argument for a landlord that there was an "unofficial tariff" of damages for discomfort and inconvenience, at that time at £2,750 per annum at the top and £1,000 at the bottom of the scale.[14] The tenant is bound to to mitigate any losses caused, as by carrying out small-scale or emergency repairs himself.[15] He cannot include a claim for outgoings on uninhabitable premises, which he might have to incur in any case,

---

[10] As in *Passley* v *London Borough of Wandsworth* (1996) 30 HLR 165 (leaking pipes in roof not let to tenant).
[11] A premium of £85,000 which a commercial tenant could have recovered from a potential assignee if the premises had not been out of repair fell within the foreseeability range in *Credit Suisse* v *Beegas Nominees Ltd, supra.*
[12] *Brent London Borough Council* v *Carmel* (1996) 28 HLR 203 at 207.
[13] *Larksworth Investments* v *Temple House Ltd (No 2), supra.*
[14] In *Wallace* v *Manchester City Council* [1998] 3 EGLR 38 at 42. See Madge "Damages for Breach of Repairing Obligations" New Law J, November 5 1999, p 1643. For recent illustrations, see eg *Gething* v *Evans* [1997] CLY 1753 and *Arnold (Paul)* v *Greenwich London Borough Council* [1998] CLY 3618 (lump sums); *Brydon* v *Islington London Borough Council* [1997] CLY 1754 (diminution in value based on the weekly rent).
[15] This requirement is not absolute: *Sturolson & Co* v *Mauroux* [1988] 1 EGLR 66, CA.

as opposed to rent that he has been forced to pay on premises which he cannot dispose of owing to their state and condition.[16]

B – Position where tenant remains in occupation

The tenant may remain in occupation despite the breaches. He may recover, as special damages for breach of the landlord's repairing covenants, his costs reasonably incurred in renting temporary alternative accommodation, if necessitated by the want of repair of the demised premises, during the period the breach of covenant continues.[17]

The tenant is entitled to recover damages to compensate him for the cost of making good any interior decorations consequent on the execution of repairs necessitated by the landlord's breach of covenant.[18] This applies even if the tenant might obtain somewhat better decorations as a result.[19] Also, he may claim for the loss of comfort and convenience of having to live in a less pleasant and habitable house than one in good repair.[20] The tenant is entitled to damages for loss of personal comfort and injury to his personal health caused by the disrepair.[21] All these items or "heads" of damages reflect the diminution in value caused by the landlord's failure to repair.

In some cases, the courts have awarded a residential tenant the difference in rent between the value of the premises in a state of repair in compliance with the landlord's covenant and the condition they in fact are in, owing to the landlord's breach of obligation.[22] The

---

[16] *Calabar Properties Ltd* v *Stitcher* [1983] 2 EGLR 46, CA.
[17] *McGreal* v *Wake* [1984] 1 EGLR 42, CA.
[18] *Bradley* v *Chorley Borough Council* [1985] 2 EGLR 49, CA (a sum of some £220).
[19] Where a landlord entered under licence to carry out works and paid the tenant £50 for redecoration, no further damages on account of injury to decorations could be recovered: *McDougall* v *Easington District Council* [1989] 1 EGLR 93.
[20] *McGreal* v *Wake* [1984] 1 EGLR 42, CA, where damages were awarded for eg the cost to the tenant of cleaning up after building works had been completed.
[21] See eg *Switzer* v *Law* [1998] CLY 3624 (£5,500 for 8 years of discomfort).
[22] See eg *McCoy & Co* v *Clark* (1982) 13 HLR 87; also *Brent London Borough Council* v *Carmel, supra*, where damages for loss of value were assessed as a percentage of the rent payable.

High Court recently made an award on this basis to a business tenant whose landlord had defaulted by not providing lift and air-conditioning services, as promised.[23]

C – Position where tenant intends sale or subletting

Where, as with a long lease, the tenant intends using the premises as an investment for sale, he will be able to recover, as damages, the difference between the sale price in fact realised with the premises out of repair, and the price which he would have been able to realise, had the premises been kept in proper repair by the landlord.[24] If the tenant intends to sublet the premises – to the landlord's knowledge – and fails to do so because of their disrepair, caused by the landlord's breach of his covenant to repair, the tenant may claim his loss of potential rental income as damages.[25]

If the tenancy is incapable of disposal on the open market, as is the case with a statutory tenancy under the Rent Act 1977, or, presumably, an assured shorthold tenancy under the Housing Act 1988, then the measure of damages is the cost to the tenant of carrying out necessary repairs (assuming the landlord refuses to do them) plus substantial damages for inconvenience and discomfort, calculated from the date of notice to the landlord until he performs his covenant or the date the damages are assessed.[26]

III – Deduction or set-off from rent

A tenant whose landlord is in breach of his covenant to repair may resort to common law and equity self-help remedies to reimburse himself for the actual or notional cost of remedial work. The

---

[23] *Electricity Supply Nominees Ltd* v *National Magazine Co Ltd* [1999] 1 EGLR 130. As there noted, a trading company cannot suffer personal losses or distress; equally it may claim for loss of profits – hence rental value diminution was there awarded, viz, the difference between: (i) the rent paid by the tenant for the premises; and (ii) their rental value in their defective condition.

[24] *Wallace* v *Manchester City Council* [1998] 3 EGLR 38 at 42.

[25] *Ibid.* Also *Mira* v *Alymer Square Investments Ltd* [1990] 1 EGLR 45, CA (damages being assessed minus sums on account of outgoings and tax).

[26] *Hewitt* v *Rowlands* [1924] WN 135, CA, which was held in *Electricity Supply Nominees Ltd* v *National Magazine Co Ltd, supra,* to be still good law.

common law remedy of set-off derives from statutes of set-off, which allow a set-off of mutual debts between landlord and tenant, in our context, provided the claims are certain or can be ascertained with certainty at the time of pleading.[27] Equity allows set-off of unliquidated, and so uncertain, claims, taking the view that, as will appear, the landlord cannot fairly claim rent if he is in breach of his obligation to repair.

A – Common law

If a landlord claims future rent but is himself in breach of his covenant to repair, express, implied or statutory, having failed to comply within a reasonable time of notice from the tenant, the latter may, having spent his own money on repairs within the landlord's express or implied covenant,[28] recoup himself out of future rent.[29] This taking of the law into the tenant's own hands[30] may carry risks. In the view of one writer,[31] if the sum spent on repairs by the tenant is excessive, it would be disallowed: thus, it may be advisable for the tenant to obtain two (independent) estimates and to submit these to the landlord or his agent prior to executing the work; and the tenant should in principle accept the lower of the two figures. Since the availability of this right depends on prior notice to the landlord, the question of whether it is available where notice cannot be given owing to emergency repairs being needed is unresolved. If the tenant decides to carry out repairs on some part of the property not demised to him, recoupment might be barred unless an implied licence to enter from the landlord can be inferred, so as to overcome the trespass which would otherwise be committed.[32]

---

[27] *Courage Ltd* v *Crehan* [1999] 2 EGLR 145 at 155, CA.
[28] In *Melville* v *Grapelodge Developments Ltd* [1980] 1 EGLR 42 at 44E, Neill J doubted, *obiter*, this particular condition.
[29] *Lee-Parker* v *Izzet* [1971] 3 All ER 1099. In *Kemra (Management) Ltd* v *Lewis* [1999] CLY 3729, set-off was limited to sums resulting from any breaches of covenant of an assignee of the reversion of an "old" tenancy, as opposed to those of his predecessor. In view of Landlord and Tenant Act 1995, s 3, no such limitation would appear to apply to tenants of "new" tenancies.
[30] Rank, "Repairs in Lieu of Rent" (1976) 40 Conv (NS) 196.
[31] Waite, "Repairs and Deduction from Rent" [1981] Conv 199 at 205; see also Waite "Disrepair and Set-Off" [1983] Conv 373.
[32] In *Loria* v *Hammer* [1989] 2 EGLR 249 an entry by the tenant onto the

B – Set-off in equity

Where rent arrears are claimed in proceedings by a landlord, and he is in breach of his covenant to repair, the tenant may invoke a general defence in equity to set off, against the amount claimed as rent arrears, any sums claimed in respect of the breaches of covenant to repair, as unliquidated damages, so as to reduce or even eliminate the landlord's claim. The sums concerned are not a fixed sum quantified in advance and so the tenant may set off the estimated costs of the repairs necessary to comply with the landlord's repairing covenant. The claim for unliquidated damages goes in equity's view to the foundation of the landlord's claim for rent arrears.[33] A claim for equitable set-off may be made by a landlord against whom breaches of repairing covenants are alleged, so as to protect him against his having to meet the tenants' claims, where the latter had not paid outstanding service charges.[34] A claim to a set-off against rent arrears may be made though the tenant has spent no actual money on repairs.[35] It is, however, assumed, equity following the law, that the equitable right of set-off depends on the tenant having given notice to the landlord of the want of repair, where required, and that the landlord has failed after a reasonable time has elapsed, to do the repairs in question.

It is open to the parties to a lease expressly to exclude the equitable right to set-off, or by necessary implication, an example of the latter being a tenant agreeing to pay rent by direct debit arrangements, since the latter are taken as the equivalent of cash payments.[36] Clear words of express exclusion are needed so that a term in an underlease requiring the sub-lessee to pay rent "without any deduction" did not suffice expressly to exclude the tenant's equitable right to set-off, which had not in terms been specified.[37]

---

[32] cont.
landlord's land to carry out emergency repairs did not deprive her of damages.
[33] *British Anzani (Felixstowe) Ltd* v *International Marine Management (UK) Ltd* [1979] 1 EGLR 65; *Eller* v *Grovecrest Investments Ltd* [1994] 2 EGLR 45, CA; *Courage Ltd* v *Crehan* [1999] 1 EGLR 145.
[34] *Filross Securities Ltd* v *Midgeley* [1998] 3 EGLR 43, CA.
[35] *Melville* v *Grapelodge Developments Ltd, supra*.
[36] *Gibbs Mew plc* v *Gemmell* [1999] 1 EGLR 43 at 50G, CA; *Courage Ltd* v *Crehan* [1999] 1 EGLR 145.
[37] *Connaught Restaurants Ltd* v *Indoor Leisure Ltd* [1993] 2 EGLR 108, CA.

Where a clause in an executed lease excludes any equitable right to set-off (as where the tenant covenants to pay rent "without any deduction or set-off whatsoever"), the test of reasonableness of section 3 of the Unfair Contract Terms Act 1977 does not apply to these clauses. This is a narrow view, based on the idea that the covenants of a lease are an integral part of the creation of a lease, creating a proprietary interest in land, and so falling within statutory exclusions.[38]

IV – Specific performance

A – *General equitable jurisdiction*

The court has an equitable jurisdiction to award specific performance of the landlord's repairing covenants. In 1996 it was claimed that specific performance was "invariably" sought in residential cases, most of these involving local authority council lettings, where the landlord had failed to carry out repairs in breach of covenant.[39] Some restricting conditions apply to this remedy.

First, the availability of specific performance is at the discretion of the court. The usual discretionary bars to relief apply. Hence, an unreasonable delay by the tenant, or proof by the landlord that damages would be an adequate remedy, would preclude an award. The work must be clear and specific. The landlord must be plainly in breach of his repairing or maintenance obligations. In addition, the order must not be futile: the landlord must be entitled to access to the property which is out of repair. If the property has not been let to the tenant, as with an exterior balcony, or with a lift, then this condition will not of itself present a difficulty. A further possible bar to relief may be that no award of specific performance can be made if the court would have to exercise constant supervision over the defendant, at least during the term of a long lease and perhaps in any case.[40] The correctness of this condition has, however, been doubted,[41] and it formed no bar to the making of an order of specific

---

[38] *Unchained Growth III plc* v *Granby Village (Manchester) Management Co Ltd* [2000] L&TR 186, CA, approving *Electricity Supply Nominees Ltd* v *IAF Group plc* [1993] 2 EGLR 95. Such a clause might infringe Unfair Terms in Consumer Contracts Regs, SI 1999 No 2083.
[39] Law Com No 238 (1996) *Responsibility for State and Condition of Property*, para 9.22.
[40] *Gordon* v *Selico Co Ltd* [1985] 2 EGLR 79 at 84H (Goulding J).
[41] By *inter alia* the Law Commission, *supra*, para 9.8.

performance of a landlord's undertaking to ensure lifts he provided functioned properly.[42]

The benefit of the availability of specific performance was shown in a case arising out of the county court. Judicial notice was taken of the fact that more determined tenants in Liverpool obtained a Scott Schedule of works, attaching it to their claim and seeking rectification of the defects and damages if these were not put right. If the landlord undertook to comply with the works detailed by the court after a period of adjournment, no order for specific performance would be made at that stage.[43] The threat of specific performance seemed to be enough to induce compliance, at least where liability was not contested.

If the item out of repair is within the demised premises, an award of specific performance in the equity jurisdiction may be impossible. The tenant may be prepared to allow the landlord a licence to gain access to the item concerned. If so, there might be no difficulty in awarding specific performance. In principle, however, an entry onto the tenant's premises to carry out repairs is, an express clause of entry apart, an actionable trespass by the landlord,[44] so ruling out specific performance where access is refused.

The operation of the inherent equity jurisdiction was illustrated by a key case[45] where the landlord was ordered by specific performance to re-instate a balcony forming part of the structure of the property in question, which was excluded from the lease, but which his landlord had expressly covenanted to repair. The inherent equitable jurisdiction is not confined to particular types of repair. The work was clear and specific and it was not oppressive to the landlord to make an order. Landlords were also ordered by specific performance to ensure that lifts they covenanted to provide and maintain, actually worked.[46] The court dismissed the landlords' objection that the order might drive them deeper into insolvency, seeing that their problems were of their own making. It will be seen that in the case of services such as providing lifts and the like, an order of specific performance is likely to be specially valuable to tenants. If a landlord ignores the court's order of

---

[42] *Francis v Cowlcliffe* (1976) 33 P&CR 368.
[43] *Joyce v Liverpool City Council* [1995] 3 All ER 110, CA.
[44] *Regional Properties Ltd v City of London Real Property Co Ltd* [1981] 1 EGLR 33.
[45] *Jeune v Queens Cross Properties Ltd* [1974] Ch 97.
[46] *Francis v Cowlcliffe* (1976) 33 P&CR 368.

specific performance, that amounts to a contempt of court. He may, for example, be ordered to pay a specific sum of money into court to cover the estimated cost of the work, which will then be carried out under the supervision of the court.

B – Statutory jurisdiction

There is a statutory jurisdiction to award specific performance under section 17 of the Landlord and Tenant Act 1985, applying only to tenants of dwellings, which expression has a wide definition. The court has discretionary jurisdiction to order specific performance of the landlord's repairing covenant relating to any part of the demised premises, whether imposed on him by statute or at common law. This provision meets the problem that under the equity jurisdiction no order is seemingly possible relating to a part of the premises demised to the tenant, as opposed to property not so demised. Under this legislation, the court would be entitled to order specific performance of a landlord's repairing covenant in relation to the whole of a common roof of, say, a block of flats.

The court expressly has jurisdiction notwithstanding any equitable rule restricting the scope of the remedy, whether on the basis of a lack of mutuality or otherwise. Section 17(2)(d) defines "repairing covenant" so as to mean a covenant to repair, maintain, renew, construct or replace any property. Hence, the statutory jurisdiction is not confined to cases of breaches by the landlord of his repairing obligations under section 11 of the 1985 Act.

C – Reform

Owing to the fact that specific performance is a means of making a recalcitrant landlord carry out the work, whereas damages do not have that result and the tenant may repeatedly have to go back to court if breaches of landlords' covenant are persistent, the Law Commission have recommended that the court should have a general power, based on a revised and expanded statutory jurisdiction, to decree the specific performance of a repairing obligation in any tenancy. The remedy would be available, stripped of the doubts and uncertainties which surround it, in any case where it would be appropriate, subject to the overriding discretion of the court.[47]

---

[47] Law Com No 238, *supra*, para 9.33.

V – Appointment of receiver or manager

If the landlord of a block of residential flats is a persistent defaulter in complying with his obligations to repair, the value of tenants' interests in the premises will be adversely affected. Hence statutory interventions in the field.

A – General jurisdiction to appoint a receiver

Section 37 of the Supreme Court Act 1981 enables the High Court to "appoint a receiver in all cases in which it appears to the court to be just and convenient to do so." The court controls his expenditure on repairs.[48] The court refused to appoint a receiver where the landlord was a local authority, because it was subjected to independent, statute-based duties, enforceable by a mandatory injunction, which were taken to give the lessees sufficient remedies.[49]

The receiver's appointment, which is likely to invest him with managerial powers in relation to the premises, may be discharged by the High Court, on a further application, by either party, if the necessity for it has come to an end. The receiver's appointment displaces the landlord from the right to receive rents: this means that the receiver will collect the rental income, providing funds, in due course, to carry out necessary repairs. Thus the court has appointed a receiver where insufficient funds existed to carry out work to flats estimated at £300,000, to collect rents and extra moneys with the assistance of the tenants.[50] It is therefore apparent, despite the width of the words "just and convenient", that the inherent jurisdiction of the court will seemingly mainly be used in cases of persistent breaches of the landlord's obligations.[51] A receiver appointed under the general equity jurisdiction cannot compel the landlord, on an interim application, to meet his expenses, where the assets turn out to be inadequate to do so, and even the expenses of drawing up a specification of repairs were held to be irrecoverable.[52] This particular problem does not apply to managers appointed under statute.

---

[48] *Boehm* v *Goodall* [1911] 1 Ch 155 at 161.
[49] *Parker* v *Camden London Borough Council* [1985] 2 All ER 141, CA.
[50] *Daiches* v *Bluelake Investments Ltd* [1985] 2 EGLR 67.
[51] *Hart* v *Emelkirk Ltd* [1983] 3 All ER 15 (flats in relation to which no rent or service charges had been collected and the premises were in a poor state of repair).
[52] *Evans* v *Clayhope Properties Ltd* [1988] 1 EGLR 33.

B – Statutory appointment of manager of block of flats

Following complaints about the mismanagement of some long leasehold flats, especially of neglect by landlords' management companies to collect service charges funds to fund repairing work on a regular basis,[53] Part II of the Landlord and Tenant Act 1987 was enacted. It has been amended by Part III of the Housing Act 1996, with a view to making it more difficult for a defaulting landlord who has been displaced from the management of the premises to regain control of their management and rental and service charge income.

The main principles of the statutory regime, which excludes within its field the inherent equity jurisdiction, are as follows. There must be leases of residential blocks of flats – purpose-built or converted. The statutory power does not apply to business tenants nor to resident residential landlords.[54] An application must be to the immediate landlord. It may be by one or more than one tenant. The leasehold valuation tribunal may order the appointment of a manager (s 24) notably if the landlord has broken a management obligation (such as to repair or to collect service charges) of the lease. The breach must be likely to continue and it must be just and convenient to make the order. Alternatively, an order may be made on the general ground that it is just and convenient to make it. The tribunal may impose such conditions as it thinks fit when making the order (s 24(6)). The manager is appointed to carry out management or receiver functions or both at the discretion of the tribunal (s 24(1)).

A preliminary notice must ordinarily be served on the landlord. This notice is in some ways akin to a landlords' statutory forfeiture notice. In particular, it must specify the fact that the tenant intends to apply for a management order, and the grounds on which an order is being requested (s 22). Specified steps must be given for the landlord to remedy the breach or breaches, within a reasonable time. The amount of detail to be given in the notice will seemingly vary with the terms of each individual lease. A landlord who

---

[53] The complaints were the basis of the Nugee Committee Report, summarised by Hawkins [1986] Conv 12.
[54] According to the government, to remove the resident landlord bar altogether might deter landlords who continue to reside in property from letting, fearing losing control of its management (Consultation Paper (2000), Part II, section 4.3, para 8).

showed no intention of complying with his obligations for some time in the past could not complain about being allowed only a short time in which to remedy his past defaults in the statutory notice.[55] Although the tenant must ordinarily wait, once a notice is served, for the expiry of a reasonable period, before seeking a management order, it appears that it is not necessary to do this if the breach concerned was irremediable, as with allowing a policy of fire insurance over the building to lapse. The fact that this is tenants' remedial legislation is emphasised by the fact that the court has a power to make a management order in the absence of a notice requirement of, notably, a reasonable period for remedy of breaches. If it is just and convenient for an order to be made without a notice, as where it is not reasonably practicable for a notice to be given, an order is still possible. This rule aims at emergency situations where urgent action is needed. In such a case it would also be possible to make an interim order pending a final order after due notice.

Once made, a management order displaces the landlord from the receipt of rents and service charges. These are colleced by the manager, who has power to effect insurance and collect premiums (s 24(11)) as well as to carry out repairs and maintenance. If, in particular, the landlord shows that, if the tribunal accedes to his application to discharge an order, the circumstances which led to the making of the order will not recur, he may be able to obtain a discharge of the order (s 24(9A)). Once a management order has been in force for two years, a gateway opens for the tenants to buy out the landlord under statutory procedures.[56]

The government, at the time of going to press, was considering reforms to the existing regime, when a suitable opportunity arose.[57] The proposed reforms would, first, allow the appointment of a manager against a third party management company. Secondly, where a majority interest in the property was held by long leaseholders, the fact that there was a resident landlord in the premises would not operate as a bar to the appointment of a manager where a majority interest in the premises was held by long leaseholders.

---

[55] *Howard* v *Midrome Ltd* [1991] 1 EGLR 58.
[56] Part III of the Landlord and Tenant Act 1987 as amended.
[57] Consultation Paper (2000) Part II, section 4.3, paras 1–9. This reform is one of those not included in the Draft Bill (*ibid*, section 1, para 7).

VI – Control of service charges

In the case of multi-occupied leasehold offices or flats, or other units, landlords may undertake obligations in relation to the repair and maintenance as well as the insurance of the main buildings and common parts and also common garden and other facilities such as garages and parking areas. The landlord may be entitled under the lease to reimburse himself for his costs by means of service charge clauses. Sometimes, long lessees are collectively liable for repairs and maintenance, and have to pay service charges to a tenants' association. If they have bought out the freehold voluntarily or under statute, the participating lessees will have granted themselves long leases: liability for repairs and maintenance would rest ordinarily with a tenants' association.

The methods of computing the liability of individual unit holders to pay their share of the cost vary. Sometimes the sums payable are a percentage of the landlord's overall costs. Equally, they may be computed on the basis of the total rateable value of the property (in the case of business premises) or the total floor area of the premises, with fractional subdivision in proportion to the size of each unit. There is a danger that some landlords will seek to carry out work to an unnecessarily high and expensive standard and so overcharge tenants, or carry out work which goes beyond their covenant to repair or maintain.[58] A further problem which has arisen is that of failure to consult lessees about major works – for whose cost they will have to pay. Other sources of dispute include such matters as not keeping up a reserve fund for major future non-recurring costs such as roof replacement costs and lack of competetive tendering by managers. In order to promote "good management practice" the RICS produced a Management Code in 1997. It is aimed not only at landlords' managers (such as an in-house management company) but at groups of flat lessees who have formed themselves into a management company. Much of its contents are common sense,

---

[58] This is not to suggest that landlords are not entitled in appropriate cases, as with long lessees, to carry out and charge for extensive works of repair if, for example, they have reasonably formed the view that the cost of longer term work is cheaper, having regard to the life-span of the premises, than less extensive work which would have to be carried out again within that life-span, as with the replacement of a flat roof with a pitch roof: see *Wandsworth London Borough Council* v *Griffin* [2000] 2 EGLR 105.

such as advice to managers routinely to monitor the cost-effectiveness of contracts, or to arrange for regular maintenance of communal space heating and gardens, where appropriate.

A – Recovery of overpayments

Service charge issues need to be considered in the light of far-reaching rulings of the House of Lords.[59] Sums paid under a mistake of law are now capable of being recovered by a payor from the payee, in contrast to the previous law, which allowed recovery only for mistakes of fact. Time runs, under the Limitation Act 1980 s 32(1)(c), from the date the mistake was discovered or could with reasonable diligence have been discovered. If the payor would have made the payment even if he had known of his mistake at the time of payment, he cannot recover the money. If the payee has changed his position after the payment has been received, because this new rule is based on restitution and so on not allowing the payee to become unjustly enriched at the payor's expense, it appears that recovery would also be barred.

These rulings may encourage tenants to press claims for recovery of overpaid service charges.[60] However, as has been pointed out,[61] there may be some limits. Not only must the tenant prove that the payment was made to the landlord or other recipient, and that there was a mistake, but it must be shown that the mistake caused the payment. Causation may not be particularly easy to prove, as where a tenant knows the law to be unclear but makes the payment in any case, taking the risk that the position may not be as he thought. Such a payor might find it difficult to recover "overpaid" service charges. If a tenant paid service charge arrears under a final court judgment, which was eventually overruled, the doctrine of finality of judgments would seem to preclude his claim, which would have to reopen the whole of the original case.

---

[59] *Kleinwort Benson Ltd* v *Lincoln City Council* [1998] 3 WLR 1095. The description "far reaching" is that of Neuberger J in *Nurdin & Peacock plc* v *DB Ramsden & Co Ltd (No 2)* [1999] 1 EGLR 15 at 20M.

[60] In *Nurdin & Peacock plc* v *DB Ramsden & Co Ltd (No 2)*, *supra*, payments of rent made on the factually mistaken assumption that a rent review had taken place were recovered by lessees, even though the tenants' error was self-induced.

[61] Miller and Pickston, "Paying for Past Mistakes", *Estates Gazette* December 19 1998.

The view has been advanced[62] that a successful claim to recovery would exist where a tenant paid a service charge thinking, at the time of payment, that he could recover the money, if the sums in fact turned out not to be due, as in later proceedings. This subjective test, if correct, is potentially restricted in its operation by the possibility of the landlord, or other recipient such as a manager, establishing a defence of change of position. This defence appears to be a general one but sadly, their Lordships did not go into much detail. It is, however, no defence for a defendant who received a payment of money under a mistake of law to prove that, before he learnt of the mistake, he honestly believed he was entitled to receive and retain the money.[63]

Since overpayments of service charges made under a tenant mistake of law can now be recovered from a landlord, it is not illogical to allow landlords who make such mistakes as underestimating the amounts of service charges in their invoices to recover, in principle, the difference between the incorrect and correct amounts, depending on the construction of an individual lease. In a recent case,[64] shop lessees in a larger precinct had to pay their landlord underpaid service charges of over £200,000. The landlord had under-assessed them for some time. The overall purpose of the lease was taken to be to allow the landlord full reimbursement of its maintenance expenditure for the benefit of all lessees in the shopping centre. The landlord's erroneous certificate could not be relied on by the lessees as conclusive. If, as was not apparently the case in this dispute, the lease had clearly provided that a certificate was final, and could not be gone behind, it is difficult to see how a landlord issuing a certificate based on a calculation he subsequently found to be erroneous could have recovered underpayments with the same ease as did the landlords in the present case. Clearly this area of the law is developing.

---

[62] Brock, "The Extent of Mistake" *Estates Gazette* March 6 1999.
[63] *Kleinwort Benson Ltd* v *Lincoln City Council, supra* at 1124–1125 (Lord Goff).
[64] *Universities Superannuation Scheme Ltd* v *Marks & Spencer plc* [1999] 1 EGLR 13, CA.

B – Reform of the law

If Law Commission proposals are ever enacted,[65] recovery for mistake of law or fact would be more restricted than at present. It would not be allowed if the payment had been made under a settled view of the law – even if that view later turned out to be erroneous. The law would be treated as "settled" if established as at the date of payment either by case law or by settled practice at that time. It is disappointing that reform of the law along these lines has not so far taken place. The Law Commission model seems to achieve a better balance between the understandable wish of lessees to recover overpayments and the policy of not encouraging stale claims, than does the present state of the law as reinterpreted by the House of Lords.

We now examine some limits which the courts have placed on the recovery of sums demanded from tenants under service charges clauses. The policy of the common law is seemingly to discourage any overcharging by landlords or managers, by confining the liability of lessees to the strict wording of the service charge clause in question.

C – Common law controls

Some general controls

Clear words in a service charge clause are construed literally, but the context of the lease as a whole governs the matter, especially if a clause is ineptly drafted.[66] The courts seek to find out the common intention of the parties, assuming that the landlord will make no profits from his services and that the tenant will be entitled to, and have to pay for, his fair share of maintenance costs. Thus, a requirement to pay service charges[67] for heating was limited to

---

[65] Law Com No 227 (1994) *Restitution: Mistakes of Law*, esp para 5.3 and Draft Bill cl 3.
[66] *Billson v Tristrem* [2000] L&TR 220, CA.
[67] If these are reserved as rent, this fact does not of itself mean that the operation of a suspension of rent clause will entail the tenant being free from paying service charges, as where the premises are partly damaged by fire, seeing that work within service charge clauses is taken to benefit the tenant for the whole term of the lease: see *P&O Property Holdings v International Computers Ltd* [1999] 2 EGLR 17.

actual costs of the landlord, who was not to make a profit.[68] Unless a clause clearly stipulates the contrary, the landlord must carry out work within its scope even if there is no fund immediately to hand.[69] Any repairing or other work which the landlord is entitled to execute must be done to a reasonable standard. No costs in respect of works of renewal are allowed if these would go outside the scope of the landlord's obligations. A landlord thus failed to recover the whole cost, £30,173, of re-roofing from office lessees. Although the roof was in disrepair, the particular service charges clause only allowed for renewal costs "if necessary". On the facts, complete renewal was not "necessary", as the roof could be mended by short-term repairs.[70] A similar approach was taken where a literal wording of parts of the lease concerned suggested that the tenant of a ground-floor flat did not have to contribute to any maintenance costs incurred by the landlord to the common parts of the premises as a whole, to which she had no access. The Court of Appeal rejected this bizarre construction, holding that the (objective) intention of the parties was that all lessees would have to pay their due proportion of maintenance costs for the whole building.[71]

Method of calculation of charges: implied terms

To prevent overcharging, there is an implied term in service charge clauses, under the guise of business efficacy, that any costs recoverable from tenants will be fair and reasonable. A landlord was not given an unfettered discretion by a service charge covenant, to adopt the highest conceivable standard of maintenance, and then to charge that to the tenant.[72] If the

---

[68] *Jollybird v Fairzone Ltd* [1990] 2 EGLR 55, CA.
[69] *Marenco v Jacramel Co Ltd* (1964) 191 EG 433.
[70] *Scottish Mutual Assurance plc v Jardine Public Relations Ltd* [1999] EGCS 43. The landlord could not, therefore, recover for works which would last for about 20 years. He recovered some 40% of the total cost, applying to short-term repairs. Contrast the result in *Wandsworth London Borough Council v Griffin* [2000] 2 EGLR 105. The evidence showed long-term cost savings for the long lessees from the landlords' repair scheme.
[71] *Billson v Tristrem* [2000] L&TR 220.
[72] *Finchbourne Ltd v Rodrigues* [1976] 3 All ER 581, CA; also *IVS Enterprises Ltd v Chelsea Cloisters Management Ltd* [1994] EGCS 14, CA.

circumstances change during the lease, so that one method of calculation becomes fair and reasonable, and an earlier one inappropriate, then the latter method can no longer be adopted. In one case, the basis of calculation for heating charges in the lease was on a floor-area basis, but this, in view of the actual area heated in the particular tenant's flat, was unfair; he was entitled to be charged only on the basis of his actual heating use.[73]

Strict approach to recovery of individual cost items

The common law does not update the terms of service charge clauses. Unless the lease expressly, or by necessary implication, enables the landlord to charge lessees with interest on borrowed money to enable him to comply with his repairing and maintenance obligations, interest cannot be recovered from tenants as part of a service charge.[74] The RICS say that this principle may, in the interests of both parties, need to be overcome by an agreed variation of the terms of the lease, since without funds, "services may not be provided".[75] Severity of approach was shown to a clause which permitted the recovery of charges of "other professional persons". It did not enable the landlord to recover legal costs incurred in trying to extract service charges from lessees.[76]

Should an item of work fall clearly within a service charge clause, it is no answer that the tenant does not want to pay or that the work might be done to a less good standard, as the landlord is not bound to carry out minimal work if this is not sufficient as a long-term remedy. Any other approach would risk damage to the long-term interests of the landlord and of all lessees. Where independent advice has been given to the landlord or manager that comprehensive work of repair is needed, a lessee was held duly liable to pay for his share of the cost.[77] In some cases, depending on

---

[73] *Pole Properties Ltd* v *Feinberg* (1981) 43 P&CR 121.
[74] *Boldmark Ltd* v *Cohen* [1986] 1 EGLR 47, CA.
[75] RICS Code, *supra*, para 9.5 (residential leases), although they admit that failure to provide services may be a breach of obligation: impecuniosity is no defence.
[76] *Sella House Ltd* v *Mears* [1989] 1 EGLR 65, CA.
[77] *Manor House Drive Ltd* v *Shahbazian* (1965) 195 EG 283; also *Wandsworth London Borough Council* v *Griffin* [2000] 2 EGLR 105.

the wording of the covenant in question, the landlord may be able to recover the cost of repairs to installations not actually situated in the particular flat of the tenant disputing the item.[78] However, the landlord cannot recover, unless clear language is used, for work contracted for, but not completed, until after the term of the lease expires.[79]

Trusts of service charges contributions

No doubt, a reputable landlord will keep all service charge payments and reserve funds in a separate bank or other account. If the terms of a lease provide that the sums paid by lessees are to be held by a management company in a fund which is to be held in a separate account, and not to be used for any other purpose than expenditure on relevant items of work, if the management company becomes insolvent, it seems that equity implies a trust at that time, so that the funds are preserved for the current tenants, subject to any claims of the landlord for outstanding work.[80] A specific statutory provision[81] applies, prevailing over any express or implied trust, where tenants of two or more dwellings are liable to pay service charges to the landlord.

Essentially, a trust is imposed by statute, seemingly from the very commencement of the lease, over the fund to apply it for the matters covered by the service charges clauses. If there is any surplus, it is to be held on trust for the contributing tenants for the time being, in shares proportionate to their respective liabilities to pay service charges. Any surplus in the fund is not to be paid to an assigning lessee. If after termination of the lease there are no contributing tenants, the fund is to vest in the landlord or his payee

---

[78] *Campden Hill Towers* v *Marshall* (1965) 196 EG 989.
[79] *Capital & Counties Freehold Equity Trust Ltd* v *BL plc* [1987] 2 EGLR 49.
[80] By analogy with *Re Chelsea Cloisters Ltd* (1980) 41 P&CR 98, CA (concerned with tenant dilapidation deposits). The general law is discussed in Hanbury and Martin, *Modern Equity*, 15th edn at p 51.
[81] Landlord and Tenant Act 1987, s 42, which does not apply to business or to regulated tenants. The government was consulting in August 2000 about reforms to the rules pertaining to accounting for leaseholders' monies (Consultation Paper Part II (2000) section 4.1, paras 1–69). They were minded to prevent fraud and to give leaseholders better information (paras 5 and 6). This is a further proposed reform not included in the Draft Bill.

(such as his agent). No doubt the idea of this provision is to protect the fund from being spent on other purposes than the maintenance of the leasehold premises and from being misapplied by the landlord or his agent to paying off debts of theirs. However, there is no sanction for a landlord who fails to comply with this provision, although it ensures that if the landlord is liquidated or goes bankrupt, the leaseholders' funds are protected from the landlord's creditors.[82]

D – Residential service charges

Control of "unreasonable" charges

Detailed legislation applies to long residential lessees whose service charges are payable to a manager or landlord. First, as from September 24 1996, any residential tenant has the right to apply to a leasehold valuation tribunal to replace the current manager of the property if he shows that unreasonable service charges have been made or are proposed or are likely to be made.[83] Under s 24(2A) of the Landlord and Tenant Act 1987, the test of unreasonableness is satisfied if the charge is unreasonable having regard to the items for which it is payable – a clear line of attack on overcharging. The test is also satisfied if the works to be paid for are of an unnecessarily high standard. Equally the jurisdiction arises if, in effect, the work is not to a sufficient standard to avoid the need for additional charges at a future date. There is also a jurisdiction to appoint a manager if it is just and convenient to do so, where none of the previously-mentioned tests apply (s 24(2)(b)).

Consultation and related rules for variable charges

Special rules as to variable service charges payable by tenants of residential dwellings and flats are in sections 18 to 30 of the Landlord and Tenant Act 1985. The legislation is aimed at landlords who have tried to make an undue profit out of service charges.[84] The detail of the legislation shows the difficulty involved in combatting this type of mismanagement. Its main aspects are here noted.

---

[82] As noted by the Consultation Paper (2000) Part II, section 4.1, para 4.
[83] Landlord and Tenant Act 1987, s 24(2)(ab).
[84] *Reston Ltd* v *Hudson* [1990] 2 EGLR 51 at 52C.

1. *Scope.* The rules apply to service charges (ie charges payable as part of or in addition to rent) imposed on tenants of dwellings, which are payable for services, repairs, maintenance, insurance or the landlord's management costs, such as employing a managing agent and where the whole or part of the charge may vary according to the relevant costs (s 18(1)). Therefore these controls do not apply if the lease provides for the payment of a service charge which is fixed at the date of the commencement of the lease. It also seems[85] that a clause providing for revision of charges in line with inflation changes as shown on a published index would escape the controls of sections 18–30 of the 1985 Act. The current controls do not apply to improvements.[86]

2. *Dual limit.* By section 19(1) of the 1985 Act, "relevant costs" are taken into account in determining a service charge payable for a period such as a year, subject to two limits. First, costs are only taken into account to the extent that they are reasonably incurred.[87] Secondly, where the costs are incurred on works or services, these latter must be provided to a reasonable standard.[88] No amount for costs incurred in excess of such standards is recoverable.[89] Thus, it was held that a service charges clause was subject to an implied right to review the basic charge (seemingly upwards or downwards) owing to the obligation of the managing company to comply with section 19(1).[90] Payments in advance of incurring the costs to be recovered are recoverable, if at all, only subject to the

---

[85] *Coventry City Council* v *Cole* [1994] 1 EGLR 63, CA.

[86] As noted in the Consultation Paper (2000) Part II, section 3, Chapter V, para 2.

[87] See *Wandsworth London Borough Council* v *Griffin* [2000] 2 EGLR 105, where landlords of blocks of flats, many of which were held by long lessees who had exercised the right to buy, were held reasonably to have replaced flat roofs with pitched roofs and windows with double-glazed windows, after advice that long-term cost savings would be achieved – however, the government wish to extend the definition of service charges to include improvements (Consultation Paper, *supra*, Section 3 Chapter V para 13).

[88] In the *Wandsworth* case, *supra*, the Lands Tribunal had no difficulty in holding that this test had been satisfied eg because the landlords had obtained six estimates from reputable contractors and had accepted the lowest of these.

[89] There is also a limit in s 20(B) in relation to costs incurred over 18 months before the landlord's demand for payment.

[90] *IVS Enterprises Ltd* v *Chelsea Cloisters Management Ltd* [1994] EGCS 14, CA.

reasonableness requirement (s 19(2)). Provision is made for adjustment after the relevant costs have been incurred. In addition, s 19(2A) allows either the tenant[91] or the landlord to apply to a leasehold valuation tribunal[92] to determine, *inter alia*, whether, among other things, costs were reasonably incurred for services, repairs, maintenance, insurance or management. The tribunal also has power to determine whether services or works for which costs are incurred are to a reasonable standard. It may be asked to rule by a tenant or landlord whether, if a service charge will be (in the future) incurred for repairs or maintenance, it would be reasonable (s 19(2B)).[93] However, as has been pointed out in the latest government consultation exercise,[94] the current definition of "service charge" does not extend to fees payable to a landlord, such as for his consent to the installation of a satellite dish, or for providing information, so leading, in the government view, to the possbility of disproportionate charges being levied.

3. *Estimates and other consultation*. The 1985 Act is taken to recognise that consultation with tenants on estimates provided to them is a matter of concern to tenants and a potential source of friction between landlord and tenant if such does not take place.[95] Where the cost of qualifying works (as defined in s 29(2)) exceeds £50 multiplied by the number of dwellings or £1,000 (whichever is the greater) the excess is disallowed unless section 20(4) or (5), discussed below, are complied with.[96] These requirements may be

---

[91] Including a tenant of what must have been an "old tenancy", who had assigned prior to the application: *Re Sarum Properties Ltd's Application* [1999] 2 EGLR 131 (but not in view of Landlord and Tenant Act 1995 s 5, an assigning tenant of a "new" tenancy). A tenant who has paid a service charge falls within s 19(2A): *R v London Leasehold Valuation Tribunal, ex parte Daejan Properties Ltd* [2000] 49 EG 121.

[92] With a right of appeal to the Lands Tribunal with leave of the leasehold valuation tribunal or the Lands Tribunal (1985 Act, s 31A(6)).

[93] In *Bounds v Camden London Borough Council* [1999] CLY 3728, it was held that a landlord who intended to carry out work to a roof before a determination under s 19(2B) could, on proper terms, be restrained by injunction from doing so.

[94] Consultation Paper Part II (2000) Section 3, Chapter V, para 4.

[95] *Martin v Maryland Estates Ltd* [1999] 2 EGLR 53.

[96] The government (Consultation Paper, *supra*, Part II, Section 3, Chapter V, para 19) proposes to replace these limits with a ceiling of a set amount, which would be revised from time to time by regulations in

dispensed with by the court if it is satisfied that the landlord acted reasonably (s 20(9)), as where emergency repairs had to be carried out to a roof and one of the lessees could be neither reasonably contacted nor notified.[97]

It appears that the £1,000 limit covers the whole of the works to which an estimate applies and it is not available for separate items of work within a works schedule. Since Parliament fails to make it clear how one batch of qualifying works is to be divided from another – a question arising where multiple works are needed – it has been held that the court must adopt a common-sense approach. Accordingly, all works covered by one contract fall within the ambit of the definition, as the latter goes to the quality of the work. Thus, landlords failed to recover the cost of additional works which had not been the subject of a separate estimate to the tenants when the need for them was discovered during works which had been estimated for.[98]

If the tenant is not represented by a recognised tenants' association then, under section 20(4), requirements about obtaining estimates and notification apply. They include a requirement of at least two estimates being obtained for the work in question, one of them from a person wholly unconnected with the landlord, plus advance notification to the tenants of impending works, and provision for the giving by the tenants and the taking into account by the landlord, of any tenants' observations.

If the tenant is represented by a recognised tenants' association, then the secretary of the association must be given a notice with a detailed specification of the works, and giving a reasonable period in which the association may propose to the landlord the names of one or more persons from whom estimates should, in its view, be obtained (s 20(5)(a)). There is a requirement that at least two estimates be obtained, one from a person wholly unconnected with the landlord (s 20(5)(b)). A copy of the estimates must go to the secretary of the association (s 20(5)(c)). All tenants liable to pay service charges and represented must also receive a notice briefly describing the works, summarising the estimates and complying with certain other requirements (s 20(5)(d)). The landlord is,

---

[96] *cont.*
the light of inflation, pointing to anomalies with the operation of the present financial limits at *ibid*, para 6.
[97] *Wilson v Stone* [1998] 2 EGLR 155.
[98] *Martin v Maryland Estates Ltd, supra.*

however, only bound to "have regard" to any observations of the tenants. Works which are not urgent cannot be begun before a date for their commencement, which must be specified in the landlord's notice (s 20(5)(g)). Under recent amendments to this legislation, a tenants' association may appoint its own surveyor to advise it on any matter relating to or which may give rise to service charges. The surveyor has wide statutory powers of entry and inspection.[99]

4. *Information about costs.* Section 21 of the 1985 Act enables a tenant, in writing, to require the landlord (or his accountant, provided he is not connected with the landlord within s 28) to supply him with a written summary of costs incurred in the last 12 months ending with the date of the request or, if the accounts are made up for 12-month periods, in the last period ending with the date of the request. In effect, full details of the costs must be provided.

Further controls

In addition to the battery of controls noted, mention should be made of section 47 of the Landlord and Tenant Act 1987, which applies to any tenant of premises consisting of or including a dwelling (s 46(1)). The landlord is required in a demand for rent or service charges to give his name and address (s 47(1)). If the address is not in England and Wales, the demand must give an address in England and Wales at which notices may be served on the landlord by the tenant. If the landlord's demand fails to comply with either or both of these requirements, the sums due under the demand are treated for all purposes as not being due from the tenant at any time before the information is furnished by the landlord to the tenant by notice (s 47(2)).[100]

Specific legislative provisions[101] allow the county court to vary, on the application of an individual long leaseholder, the service

---

[99] Housing Act 1996, s 84 and Sched 4.
[100] Section 48 of the 1987 Act requires the landlord, on pain of irrecoverability, to serve on the tenant a notice giving a name and address for service on the landlord of tenants' notices. An unqualified statement of the landlord's name and address in the lease or tenancy agreement sufficies for the purpose, although, if the landlord were to move, a further s 48 notice would then be required: *Rogan v Woodfield Building Services Ltd* [1995] 1 EGLR 72, CA.
[101] Landlord and Tenant Act 1987 Part IV ss 35–39. In August 2000, the government consulted as to long-term reform of these provisions,

charge provisions of long leases which fail, in particular, to make satisfactory provision for repairs or maintenance of the flat, the building in which the flat is contained or land and building let to the tenant under the lease, as to the computation of service charges, or for the insurance of the flat, or for computation of service charges under the lease. In the case of a long lessee of a dwelling, however, the right to make an individual application to vary the lease terms is confined to unsatisfactory insurance provisions in the lease concerned.[102] These provisions exist alongside the ability of the parties to a lease to agree on contractual variations to their terms, but the government think that an advantage of the legislation is that it avoids the tenant paying money to the landlord in return for a variation,[103] which may explain their proposals to extend the scope of these provisions.

A concluding note on reform

The government was consulting, at the time this book went to press, about further reforms in the field of service charges. In addition to those already mentioned, the government floated, *inter alia*, the idea of imposing a requirement of consultation with tenants on long-term maintenance contracts. Such would include a contract to appoint a managing agent and as to annual insurance cover. No minimum sum would be imposed for such contracts since the government thought that some landlords deliberately manipulated work programmes so that no item triggered any requirement to consult. The sanction for non-compliance would be total irrecoverability of all costs incurred under the contract concerned.[104]

---

[101] *cont.*
which they think are too narrow (Consultation Paper (2000) Part II, Section 4.2, paras 1–18). They propose eg to clarify the existing grounds for an application and also propose to allow leases to be varied to permit advance payment of service charges. Jurisdiction would be transferred to leasehold valuation tribunals.
[102] Landlord and Tenant Act 1987, s 40.
[103] Consultation Paper (2000) Part II, Section 4.2, para 3.
[104] *Ibid*, Part II, Section 4 Chapter V, paras 21 and 22, and draft Bill clause 113, which would revise the whole of s 20 of the 1985 Act. The government was also proposing to transfer jurisdiction as to both failure to consult and as to reasonableness of service charges to leasehold valuation tribunals (para 26). These proposals have high priority (*ibid*, section 1, para 7).

A more radical, high priority reform, would be the enactment of a leaseholder's right to manage. The scheme in the form proposed in August 2000 is detailed, as shown by its taking up 23 clauses and one Schedule in the (draft) Commonhold and Leasehold Reform Bill 2000. The right to manage would provide, the government says "an alternative option for leaseholders who are dissatisfied with the management of the building".[105] If leases were very long, or leaseholders wished to try managing the building themselves before committing themselves to buying the freehold, the right to manage would be available. The machinery would resemble the procedures, which the government proposes to amend, for the collective enfranchisement of the freehold reversion by qualifying long lessees.

---

[105] Consultation Paper (2000) Part II, Section 2.1, para 2. The details of the proposals are in Part II, Section 3, Chapter I.

Chapter 9

Insurance and reinstatement of premises

I – General principles

The lease may provide that the tenant is not liable under a covenant to repair to rebuild the whole or any part of the demised premises in the event of their destruction by fire or other insurable risk such as storm, flood, tempest, subsidence, earthquake, landslip or other Act of God, or even damage caused by inherent defects. The landlord may expressly covenant to rebuild the whole or any part of the demised premises destroyed by fire or other insured event. To enable the cost of the necessary work to be met, the landlord may covenant to insure the whole of the premises to their full reinstatement or replacement value or some other equivalent against fire and other insurable risks. If the tenant rather than the landlord is liable to take out insurance, the policy will be only to the extent of the former's interest in the premises, so that if these are multi-occupied, a block insurance taken out by the landlord's name or in the joint names of the parties may be used. A covenant to insure by either party must be effected within a reasonable time of the covenant coming into effect and is broken if for any length of time the premises are uninsured.[1]

The necessity for an express covenant to insure arises from a principle that the landlord cannot be sued for failure to insure where he has not expressly covenanted to do so and, if he has insured, he is not impliedly liable for a failure to renew a policy. Where landlords who were not expressly bound to insure had taken out, but not renewed, a policy of block insurance, a loss of some £70,000 to the tenants resulting from a fire which was not insured against was irrecoverable from the landlords.[2]

---

[1] *Penniall v Harborne* (1848) 1 QB 368. A breach of a tenants' covenant to insure is single, not continuing, for forfeiture purposes: *Farimani v Gates* [1984] 2 EGLR 66, CA

[2] *Argy Trading Development Co v Lapid Developments Ltd* [1977] 1 WLR 444.

The landlord may be required by the lease to produce a copy of the insurance policy on demand to the lessee and to notify him of any receipt of insurance moneys. In the case of dwellings and flats, the tenant is entitled, where he has to pay service charges[3] for or towards the cost of landlords' insurance premiums, to specific information so as to enable him to discover what the insurance cover is, inspect the insurance policy, and entitling him to notify the insurer of any possible claim.[4]

The selection of an insurance office (sometimes in terms "reputable")[5] may be that of the landlord, even if the tenant is responsible for taking out insurance. There is an obvious danger to the tenant of having to pay inflated insurance costs. In the absence of express words in a lease such as that the tenant is only to pay reasonable sums to cover the landlord's insurance costs, the tenant's protection against overcharging, where he has to reimburse the landlord's insurance costs, lies in the requirement that "the landlord cannot recover in excess of the premium which he has paid and agreed to pay in the ordinary course of business as between the insurer and himself. If the transaction was not arranged in the ordinary course of business ... then it can be said that the premium was not properly paid, having regard to the commercial nature of the leases in question".[6]

However, applying the principle that a term will only be implied into a contract if it cannot work without that term, no term was implied into the covenants between a flat management company and a landlord. Using a power to do so in the lease, he required the management company to use a new and more expensive insurer. A term that the insurance premium payable should not be unreasonable or alternatively that the tenant could not be required to pay a substantially higher sum than the tenant could himself

---

[3] Owing to the definition in s 18(1) of Landlord and Tenant Act 1985 of "service charge", sums payable as additional rent on account of insurance paid by the landlord appear to fall within that Act.
[4] Landlord and Tenant Act 1985, Sched.
[5] Ross, 4th edn, Appendix 1 Form 1.1., cl 12.1.2; cf Ross, 5th edn, Precedent 1, cl 5.3.1. ("substantial and reputable").
[6] *Havenridge Ltd* v *Boston Dyers Ltd* [1994] 2 EGLR 73 at 75M (Evans LJ), CA. His Lordship added, *ibid*, that an application to the parties of the "officious bystander" test would lead to the same result.

arrange to pay a reputable insurer was rejected.⁷ There was also held to be no implied obligation on a landlord to "shop around" for the cheapest insurance. No obligation to do so could be spelt out of a requirement in a lease that the insurance had to be with "some insurance office of repute".⁸ It has been claimed that a landlord must exercise his power of nomination of an insurer in good faith,⁹ since he is spending someone else's money – not much of a safeguard, however, given the difficulty of proving bad faith.¹⁰

The principles respecting insurance and reinstatement seem to apply with equal force as between freeholders and long leaseholders and their mortgagees as they do between landlord and tenant. So as to preserve their security, mortgagees sometimes insist that it is a condition of the mortgage of individual residential or commercial units that the mortgagor is to keep down a proper policy of fire, etc, insurance, and, in the case of a block of flats, that a block insurance policy is taken out by the lessor. A failure to take out or to keep down such a policy would presumably allow the mortgagee to enforce his security, as by taking possession, even if the mortgage instalments were being promptly repaid.

II – Further considerations

A – Basis of insurance

It has been said that "insurances on property are *prima facie* to be construed as contracts of indemnity. Subject to the express terms of the policy the measure of the indemnity is the diminution in value of the thing insured as a result of the insured peril."¹¹ Where a landlord had insured the premises in his sole name, and received

---

[7] *Berrycroft Management Co Ltd* v *Sinclair Gardens Investments (Kensington) Ltd* [1997] 1 EGLR 47, CA. The court reserved its opinion as to whether para 8 of the 1985 Act, Sched, might apply in favour of the lessees had the tenancy in terms given the landlord the right to nominate the insurer, which on the facts it did not do.

[8] *Havenridge Ltd* v *Boston Dyers Ltd*, *supra*; see further Tromans, 13–05.

[9] Ross, 4th edn, para 10.15 (since the landlord is spending the tenant's money).

[10] For powers of the county court to vary any insurance provisions of long leases of flats and other dwellings, see Landlord and Tenant Act 1987 Part IV (ss 35–40).

[11] *Lonsdale & Thompson Ltd* v *Black Arrow Group plc* [1993] 1 EGLR 87 at 89M.

an indemnity in full for their loss after a negligently caused fire, he failed to recover damages from the tenant, as this would be to give him double compensation.[12] The tenant in question had been paying for the cost of the insurance and the court inferred from the facts that the policy was for the joint benefit of the parties. Where a landlord's insurance is the joint names of the parties, it is presumed that the policy covers the full extent of his and the tenant's interest; and both parties may then hope to be informed by the insurance office of any danger of lapse of the policy.[13] No doubt one aim of an indemnity from insurance moneys is that reinstatement of the buildings is to take place – if clearly so stated including any buildings to be put up during the lease.[14] Such will be aimed at without the risk of a shortfall – in which latter event the party liable to reinstate must, in the absence of a term in the lease limiting his obligation, find the additional costs out of his own pocket.[15]

Insurance policies may provide for various bases of insurance, such as "full market value", or for "full reinstatement cost" or "full reinstatement value" or the "full value" of the building or premises.[16] The general principle of such formulae seems to be to

---

[12] *Mark Rowlands Ltd v Berni Inns Ltd* [1985] 2 EGLR 92, CA (special facts since the tenant had eg relief from his obligation to repair in the event of fire, the landlord being under an obligation to reinstate in that latter event). Where, as a matter of construction, the insurance was for the landlord's sole benefit, the tenant had no defence to an action by the landlord in negligence on account of a fire: *Lambert v Keymood Ltd* [1997] 2 EGLR 70. Cf *Independent Tank Cleaning v Zabokrzeki* (1997) 8 RPR (3d) 177 (Ont Sup Ct), where the fact that the tenant was paying for increases in premiums caused to his landlord if his business increased the fire risk led to the inference that the relevant insurance policy was held for the joint benefit of both parties, even though the landlord was not expressly bound to insure the premises.

[13] Ross, 5th edn, para L229. A lapse would occur if a premium went unpaid; but as to the question whether there would be any lapse on account of radically changed circumstances not notified to the insurance office see *Kausar v Eagle Star Insurance Co Ltd* The Times, July 15 1996, CA.

[14] The landlord is not liable, in the absence of clear words, to replace tenants' improvements, "as he might be ruined in many cases": *Loader v Kemp* (1826) 2 C&P 375.

[15] *Digby v Atkinson* (1815) 4 Camp 275.

[16] See eg *Precedents for the Conveyancer*, Nos 5–8 (cl 4(2); 5–41 (cl 6(3)); 5–42 (cl 5(2)); 5–60 (cl 5(iv)).

provide the landlord (or tenant as the case may be) with sufficient funds from the insurance policy to enable him to rebuild the premises at the costs prevailing at the time of the insured event, thus taking due account of inflation in building costs. If the policy is expressed to be held in the joint names of the parties, or if the court as a matter of construction implies that it is, as where the tenant is under an obligation to pay for or contribute to the cost of insurance, or the landlord is required to reinstate the premises, or there is an abatement of rent provision applying during reinstatement, then the landlord has no right to recoup the cost of reinstatement from the tenant: neither has his insurance office.[17]

The landlord may wish to include in the insurance cover a sum in respect of loss of rent over a period of time to be agreed. Although, at common law, the tenant remains liable for rent despite the fact that the demised building may have been destroyed or rendered unusable by fire or other insured event,[18] and the lease would not be frustrated provided the underlying land was not wholly destroyed,[19] the lease may expressly provide that the rent is subject to a cesser, suspension or abatement of rent clause. Under such a clause, the tenant might be wholly or partly freed from his liability to pay rent for a specified period as from the loss or destruction of the relevant building, or until its reinstatement, whichever period is the shorter. If the building concerned is damaged but still useable in part, there may be provisions for a reasonable abatement of rent as determined by a surveyor.[20]

B – Reinstatement of building

If the landlord is under an obligation to insure, he may also be subject to an express obligation to reinstate the insured building after a fire or other insurable event. A landlord's covenant may contain an express limitation that he is not liable in any event to reinstate if the tenant or any person for whom the tenant is responsible (such as his employee or licensee) vitiates the insurance policy by his act or default.

---

[17] See Heller and Kavanagh, "The Tenant's Nightmare" *Estates Gazette*, December 5 1998, p 151.
[18] *Redmond v Dainton* [1920] 2 KB 256; *Matthey v Curling* [1922] 2 AC 180.
[19] *National Carriers Ltd v Panalpina (Northern) Ltd* [1981] AC 675.
[20] As in eg *Precedents for the Conveyancer*, No 5–8, cl 4(3) and 4(4).

The covenant may expressly require a landlord to rebuild with reasonable diligence once the insurance moneys are received. Since planning, building regulations and other consents for any rebuilding may be required, the lease may qualify the landlord's obligation to reinstate by providing that he is not bound to reinstate if, despite all his reasonable endeavours to obtain any necessary consents, reinstatement is impossible because such consents have been refused. Under one scheme, the landlord is also entitled to notify the tenant that reinstatement is not in fact possible. Thereupon the tenant has a right by notice to the landlord, served within a specified time, to surrender the lease.[21] Depending on the terms of any such clauses, the tenant may or may not be entitled to any of the insurance moneys. The parties may agree not to reinstate the building concerned, as where it is redundant or not suitable for any present use, or reinstatement may be impossible or impracticable. To cover such contingencies, the lease may expressly provide for the division of the insurance moneys between the parties on an agreed basis or, in default, as determined by arbitration in specified proportions.[22]

It is said that "express provision is better than reliance on the authorities".[23] If, however, the parties agree only that full reinstatement is not going to take place but the lease makes no provision as to the division of insurance moneys, much may depend on the facts, the circumstances and the conduct of the parties. In one case, it was held by a divided Court of Appeal that a lessee, who was under an obligation to pay for the insurance and to reinstate, was entitled to all the insurance moneys where reinstatement was impossible owing to the making of a compulsory purchase order. It did not matter that the policy was held in the joint names of the parties, at least according to the majority of the court, since the joint names requirement was a means of making sure that the premises would be reinstated using the insurance moneys.[24]

Following this approach, if the landlord carries the cost of the insurance, and reinstatement is not possible, he should be entitled to keep all the moneys. However, the parties may agree expressly

---

[21] *Precedents for the Conveyancer* No 5–58.
[22] See eg *Precedents for the Conveyancer* 5–10 (cl 12(3), proviso).
[23] JE Adams [1984] Conv 404, 405.
[24] *Re King* [1963] Ch 459, CA; WG Wellings (1964) 28 Conv (NS) 304.

or even by conduct to value each party's interest and to share out the moneys in proportion thereto, which may seem fairer since the tenant has only a limited interest in the premises. In one case, the landlord insured premises in the joint names of the parties and was liable to reinstate, but where reinstatement of the premises after a fire was unrealistic due to a large shortfall in the insurance funds, the parties were taken to have impliedly agreed to divide the insurance moneys between them in shares proportionate to the value of their interests in the building, valued at the date of the fire. The tenant could not insist on damages for failure fully to reinstate. Such a requirement was inconsistent with both parties' conduct: thus, the tenant had abandoned the premises and had unconditionally released half the moneys to the landlord, who thereafter did not keep down the insurance policy.[25]

The reinstatement of a building to the same or, as may well happen, to substantially the same state and condition as prior to the event[26] may not be practicable or useful, as where the premises are used for business purposes, and were old when let, and where a new viable commercial unit could easily be much smaller and built to a different design. In such a case, an obligation to reinstate the premises might be expressly agreed to enable rebuilding by the landlord to an appropriate equivalent of the destroyed building. Where reinstatement however does take place, questions arise as to whether the same materials are to be used in rebuilding the interior as before. In relation to interior finishes and floorings, a slavish reconstruction of what was there before is not inevitably required, but the intended or permitted use of the premises after reinstatement seems relevant. If it is to be the same as before, the courts may be suspicious of a proposed substitute material offered by the landlord. A tenant was thus entitled to reject a proposed new linoleum floor as a replacement for a destroyed wooden floor in a building whose continued use for making high-quality clothing was envisaged, even though the advantages of the latter were not much greater than those of the former.[27] The tenant of a building used as a theatre would, but for estoppel, have been entitled to insist on exact replacement of the decorations, and so on the use of gold leaf decorations on the ceilings after a fire. It might well have

---

[25] *Beacon Carpets Ltd* v *Kirby* [1984] 2 All ER 726, CA.
[26] As to which, see *Times Fire Insurance Co* v *Hawke* (1858) 1 F&F 406.
[27] *Vural Ltd* v *Security Archives Ltd* (1989) 60 P&CR 258.

been different if the intended use after reinstatement had been radically different, for example, as a warehouse,[28] where work of that latter standard would seem pointless or useless to either party.[29]

C – No express covenant to reinstate

In the, now perhaps relatively unlikely, absence of an express or implied[30] obligation by the landlord to reinstate in the event of fire or other insured peril, section 83 of the Fires Prevention (Metropolis) Act 1774 (whose application is general and not confined to London) enables any person interested in any "houses or other buildings", including the lessee or sublessee or a mortgagee, to require to the insurance office, before they have paid out the insurance moneys, to apply the funds in reinstatement of the premises. The 1774 Act has limits: it does not apply if a fire has been started intentionally or where it has been caused by the negligence of the tenant.[31] In addition, by referring to "houses and buildings", it necessarily excludes tenant's trade fixtures, removable by him at the end of the tenancy, since if the landlord had conveyed his reversion immediately before the fire, these items would not have passed.[32]

If a landlord, tenant or a mortgagee wishes to compel an insurance office to spend insurance moneys on reinstatement, his remedy is evidently in mandamus – not in the form of a mandatory injunction.[33] Where neither landlord nor tenant could agree as to the manner of reinstatement, the best remedy was said to have been an injunction to restrain the insurance office from paying the moneys to the party who had effected the insurance without giving sufficient security.[34]

---

[28] *Camden Theatre* v *London Scottish Properties Ltd* (Unrep) November 30 1984, cited in Woodfall, 11.105 n.29.
[29] Woodfall, 11.105.
[30] As to which see *Mumford Hotels Ltd* v *Wheler* [1964] Ch 117, where a lessor's obligation to insure was held on the facts to require reinstatement.
[31] *Musgrove* v *Pandelis* [1919] 2 KB 43.
[32] *Ex parte Gorely* (1914) LJ Bkcy 1.
[33] *Simpson* v *Scottish Union Insurance Co* (1863) 1 H&M 618.
[34] *Wimbledon Park Golf Club Ltd* v *Imperial Insurance Co Ltd* (1902) 18 TLR 815.

Chapter 10

Liability to repair imposed by statute

I - Introduction

Although the traditional method for regulating liability for repairs and maintenance in a lease is by using express covenants, there are a number of specific instances where Parliament has imposed repairing obligations on the parties to a lease. The most important of these is the statutory rules benefitting short residential tenants, who are a vulnerable group. They cannot be expected to have the financial means, or the incentive, to carry out any but the most small-scale repairs to the house or flat they occupy, damage caused by their own fault excepted. Statute has also imposed specific repairing or fitness obligations accross a wider field than that of short leases, and these duties, governing such diverse fields as long lessees of flats, owners of unfit housing or of property which constitutes a statutory nuisance, are noted in this Chapter. Even in the business tenancy renewal sector, statute provides a rent-fixing formula which may allow the courts to discount the full market rent if the premises are dilapidated: we note these principles.

For all their importance, the statutorily implied covenants to repair are limited in scope.[1] The landlord is not obliged to go beyond the statutory implied covenant to keep in repair so as to improve the premises or to make them fit for habitation, if they are in repair,[2] which can happen if an item is undamaged but unfit for use (such as with metal windows prone to severe condensation or a wooden door which is undamaged but not draught-proof).[3] This

---

[1] *McGreal* v *Wake* [1984] 1 EGLR 42, CA; also *London Borough of Newham* v *Patel* (1978) 13 HLR 77, CA.
[2] *Quick* v *Taff-Ely Borough Council* [1985] 2 EGLR 50, CA.
[3] While in *Windever* v *Liverpool City Council* [1994] CLY 2816 (Cty Ct) "off-level" floors were deemed not to be in repair, as not performing the function they were designed to perform, this wider formulation of liability is hard to reconcile with dicta in *Stent* v *Monmouth District Council* [1987] 1 EGLR 59 at 64, CA. However, the floor in its then state

is a lacuna, in view of the redundancy of section 8 of the Landlord and Tenant Act 1985, and because section 4 of the Defective Premises Act 1972, imposing tort liability on a landlord for certain dangerous defects, is firmly linked to the covenant to repair.[4] The implied statutory obligation has been complied with by short-term running repairs, even if a long-term solution would call for more extensive work.[5]

If a landlord in a particular tenancy agreement chooses not only to follow the wording of the implied statutory obligation, but also to add additional express words, he will be bound by these, since all words in a covenant are given their proper meaning. Hence a local authority landlord which expressly undertook to keep a ground-floor flat in "good condition and repair" was held liable in damages to the tenant, for not having put right a lack of proper insulation. In its defective state, the flat suffered from excessive condensation and severe black spot. A lay person would presumably think such premises to be out of condition.[6] Although it would be unwise to put too much weight on the result in one case turning on particular words, this result shows that if landlords choose to go beyond the wording of the statutory obligation, they do so at their own peril, having regard to the vulnerable status of social tenants and because the courts are seemingly beginning to chafe under the artificial limits imposed on landlord liability by the statutory implied obligation.

Reforms have been proposed by the Law Commission. They proposed a complete code, built up from strands of the present patchwork of common law and statute, for short residential tenancies, consisting of the current statutory repairing obligations, and an implied covenant by the landlord that the premises are to be and kept fit for human habitation. In addition, so as not to

---

[3] *cont.*
could be said to have been physically damaged, thus triggering landlord liability.
[4] *McNerny* v *Lambeth London Borough Council* [1989] 1 EGLR 81, CA.
[5] *Dame Margaret Hungerford Charity Trustees* v *Beazeley* [1993] 2 EGLR 143 (short-term patch repairs to a roof sufficed on facts).
[6] *Welsh* v *Greenwich London Borough Council* [2000] 49 EG 118. Account was expressly taken of the fact that the tenancy was in a short and simple form and that the landlord was a provider of social housing, so that the agreement must be construed in the same way that an ordinary person in the street would have construed it.

encourage tenants to be bad tenants, an implied covenant would be imposed on the tenant to take proper care of the premises.[7] For reasons given in relation to local authority unfitness controls, which may face a major overhaul, it is not at present clear whether the useful Law Commission reform package will be enacted in the form proposed.

II – Statute-implied repairing obligations of landlords of short residential tenants

Sections 11 to 16 of the Landlord and Tenant Act 1985 imply obligations to repair in the case of leases of "dwelling-houses" (ie houses, flats and appurtenances) on the landlord. The statutory duty applies to terms for less than seven years and so to assured shorthold tenancies under the Housing Act 1988, to protected or statutory tenancies under the Rent Act 1977, to secure tenants,[8] and also, it appears, to a person holding a "non-proprietary lease" from a landlord who himself holds no estate in the land from his own lessor.[9]

If an initial term certain falls within the Act and it continues, after expiry, from year to year, it is presumed that the 1985 Act would apply to the implied periodic tenancy, even if the total length of the term granted regarded back from the date of original grant of the fixed term to the eventual date of expiry after notice to quit of the implied tenancy following exceeded seven years. Periodic implied yearly tenancies are terminable by notice on either side and if neither side serves notice on the other, he consents to a renewal of the tenancy only for a further period of one year at a time.[10]

The effect of the statute is to imply the relevant obligations against the landlord into the relevant tenancy. The obligations are

---

[7] Law Com No 238 (1996), para 7.13.
[8] Including a secure tenant who has had an order for possession made against him discharged, whereupon the statutory obligations of the landlord are retrospectively revived: *Lambeth London Borough Council* v *Rogers* [2000] 1 EGLR 28, and also a former secure tenant, known as a "tolerated trespasser", against whom an order for possession has been varied: *Pemberton* v *Southwark London Borough Council* [2000] 2 EGLR 33, CA.
[9] *Bruton* v *London & Quadrant Housing Trust* [1999] 2 EGLR 59, HL.
[10] *Hammersmith and Fulham London Borough Council* v *Monk* [1992] 1 AC 478 at 490 (Lord Bridge).

contractual in nature.[11] Thus, in relation to physical damage occurring within the confines of the demised premises, notice is, exceptionally, required to the landlord of the defect before he is liable to undertake the work.[12] The usual contractual remedies, however, become available to the tenant for breach by the landlord of his statutory obligation. It appears that specific performance is a key remedy in this field in county courts.[13] Moreover, an assured shorthold tenant whose landlord refused to comply with his statutory repairing obligations in respect of serious breaches rendering part of the house concerned uninhabitable was held entitled to treat these breaches as going to the root of the contract of tenancy and as amounting to repudiation by the landlord of the agreement.[14] The tenant was entitled to leave the premises, was freed from future liability for rent and was able to claim damages from the landlord in repect of matters such as lack of heating during the winter, a collapsed ceiling, and incursion of water into the sitting-room.

A – Application of statute-implied duty

General rule

The statutory obligations, which have been said to be of enormous social importance, and to have done much to encourage repairs,[15] apply to a lease of a dwelling-house granted on or after October 24 1961, for a term of less than seven years. To avoid artificial backdating of leases to evade the 1985 Act, the period of less than seven years is computed as from the date of grant of the lease (ie in the case of a lease by deed, its delivery) which might differ from the stated commencement date.[16] It is presumed (s 13(2)(b)) that a landlord who grants a term above the statutory period but with an unfettered right or option in the lease to terminate it within seven years will exercise that right, so that the Act applies to the lease as from the outset. By contrast, it is assumed (s 13(2)(c)) that a tenant who has a tenancy for less than seven years initially, with a

---

[11] *O'Brien* v *Robinson* [1973] AC 912 at 927G (Lord Diplock).
[12] *British Telecom plc* v *Sun Life Assurance Society plc* [1995] 2 EGLR 44, CA.
[13] Law Com No 238, para 5.12.
[14] *Hussein* v *Mehlman* [1992] 2 EGLR 87.
[15] Law Com No 238, para 5.12.
[16] *Brikom Investments Ltd* v *Seaford* [1981] 2 All ER 783, CA.

common law option for renewal of the lease beyond that time, will exercise his right and so the Act is excluded from the outset. An option to terminate is distinct from a proviso in the tenancy enabling either party to determine it if the landlord dies during the term, so that if the initial duration of the tenancy in this case is for over seven years, the Act will not apply to it.[17]

"Lease" includes a head or a sublease (s 38(2)(a)), so that while a head landlord may grant a term for a length outside the 1985 Act, a sublease of the whole or part of the premises, if let wholly or mainly as a private residence (s 16(b)), would be caught.

Exclusions

Specifically excluded from section 11 are certain types of leases (s 14(4)), notably leases granted as from October 3 1980 to a local authority or registered housing association: these bodies are not in the same vulnerable position as a private individual tenant. A new tenancy granted to an existing business tenant within Part II of the Landlord and Tenant Act 1954 is not subject to section 11 of the 1985 Act, even where the 1954 Act Part II has been contracted out of in relation to the previous lease (s 32(2)).[18] The statutory obligations cannot be contracted out of in the tenancy itself, either by words to that effect or by charging the tenant with any part of the cost of repairs falling on the landlord (s 11(4) and (5)).[19]

Where obligations are imposed on a tenant in a tenancy to which the 1985 Act applies, the implied obligations of the landlord must be subtracted from any express repairing and decorating obligations of the tenant to see what obligations, if any, the latter has once the landlord has fulfilled his own obligations.[20] Thus, if in a tenancy agreement, the tenant has undertaken liability for internal decorations, but the landlord has to repair an external wall, owing to section 11 of the 1985 Act, and in the process damages the interior decorations, since the courts, after some hesitation, have concluded that he must make good the latter,[21] the

---

[17] *Parker* v *O'Connor* [1974] 3 All ER 257, CA.
[18] The Crown is not within s 11 of the 1985 Act: *Department of Transport* v *Egoroff* [1986] 1 EGLR 89, CA.
[19] The county court may order the inclusion in the lease of a term modifying the statutory covenants (s 12).
[20] *Irvine* v *Moran* [1991] 1 EGLR 261 at 262C.
[21] *Bradley* v *Chorley Borough Council* [1985] 2 EGLR 49, CA.

subtraction process referred to would leave the tenant with no liabilities on that occasion.

B – Duty to keep structure, exterior and installations in repair or working order

General duty

The 1985 Act implies a covenant by the landlord into a short residential tenancy to "keep in repair the structure and exterior of the dwelling-house" (s 11(1)(a)). Drains, gutters and external pipes are expressly included. The obligation of the landlord does not, in contrast to the duties implied by section 8 of the 1985 Act, extend to the whole of the "house" concerned.[22]

The standard of repair required is not laid down. Regard is to be had, in determining the standard of repair, to the same matters, ie the age, character and locality of the premises, as are applicable to short leases at common law,[23] with the specific addition of the prospective life of the "dwelling-house" (s 11(3)). Thus a standard applying to long leases is not thought appropriate for what is taken as short-term accommodation benefitting in the main from short-term repairing work. The general interpretation of the words "structure and exterior" has been mentioned elsewhere (Chapter 4). The structure of a dwelling-house is something less than the overall dwelling-house, so limiting the scope of the landlord's obligations. Hence it was said:

> The structure of a dwelling-house consists of those elements of the overall dwelling-house which give it its essential appearance, stability and shape. The expression does not extend to the many and various ways in which the dwelling-house will be fitted out, equipped, decorated and generally to be made to be habitable.[24]

In the case of a house, as opposed to a flat, the extent of the landlord's obligations is determined by the extent of the demise in each case and is a question of fact. If the landlord retains in his control any parts of the premises, he is not liable under statute to keep these in repair. He may be liable in nuisance.

---

[22] As noted by the Law Commission Report (1996), para 5.20.
[23] *Proudfoot* v *Hart* (1890) 25 QBD 42, applied in *Jaquin* v *Holland* [1960] 1 All ER 402, CA.
[24] *Irvine* v *Moran* [1991] 1 EGLR 261 at 262F.

Liability to repair imposed by statute

The landlord might be under an implied obligation to repair if the item concerned is an essential means of access such as a path, without which the house cannot be used, but which is not demised to the tenant or within the scope of the statutory duty. It was held extravagent and unsupportable to say that a path running between certain other nearby premises in a council estate fell within section 11 of the 1985 Act.[25] As a matter of fact and degree, a path at the front, but not one at the rear, of a council house formed part of the "exterior"of the house within the statutory duty.[26] The landlord of a house (or flat) is liable to keep a window in repair if it is external (outward-facing) or structural, in the event of the structure being made up, in part, of windows, provided the item is in a condition of disrepair rather than being unfit for use, for reasons given. It was conceded that internal plasterwork, if damaged, fell within a landlord's implied duty,[27] so that where a house suffered from dampness caused by saturated and progressively deteriorating plaster, the landlords were liable in general and special damages. However, in an earlier case, the view was expressed that internal wall plaster was more in the nature of a decorative finish, not part of the "essential material elements that go to make up the structure of the dwelling-house",[28] but this view seems to have been impliedly overruled by the case previously mentioned. Moreover, it is difficult to see how a house whose plasterwork is in such a poor state as to cause material damage to the rooms could be said to be structurally in repair, even though a house with plaster in that state may not be in danger of collapse.

Duty of landlord as to installations

The landlord of short residential lessees is also under a statutory obligation "to keep in repair and proper working order the installations in the dwelling-house for the supply of water, gas and electricity and for sanitation (including basins, sinks, baths and sanitary conveniences, but not other fixtures, fittings and appliances for making use of the supply of water, gas or electricity)"

---

[25] *King* v *South Northamptonshire District Council* [1992] 1 EGLR 53 at 54K.
[26] *Brown* v *Liverpool Corporation* [1969] 3 All ER 1345, CA; *Hopwood* v *Cannock Chase District Council* [1975] 1 WLR 373, CA.
[27] *Staves* v *Leeds City Council* [1992] 2 EGLR 37, CA.
[28] *Irvine* v *Moran, supra* at 262M.

(s 11(1)(b)). The landlord is also required to keep in repair and proper working order the installations in the "dwelling-house" for space heating and heating water (s 11(1)(c)). By statutory instrument, the landlord must ensure that a gas appliance such as a cooker or heater, installed in the premises or any part by him is maintained in a safe condition so as to prevent the risk of injury to any person. To that end he must see to it that any such appliance is checked for safety at minimum intervals of 12 months by an approved person.[29]

The duty to keep in repair "and proper working order" requires the landlord to carry out work to ensure that the physical or mechanical condition of the installation is kept in a state sufficient to enable it to function properly as such.[30] Thus, if a water cistern floods the house or flat because it was designed defectively, the landlord is liable both to replace it and in damages for consequential losses to the tenant.[31] The landlord is not bound under the statutory covenant to improve an installation, so that he is not required by the Act to lag water pipes if these were not previously lagged.[32]

Extended obligation in the case of flats

In the case of a short lease of a flat, the above obligation is extended, in the case of a lease entered into as from January 15 1989, to "an installation which, directly or indirectly, serves the dwelling-house" (s 11(1A)(b)). As with the extended obligation to repair the structure and exterior of flats, discussed below, this duty only arises if the disrepair affects the tenant's enjoyment of his own flat or the common parts he is entitled to use (s 11(1B)).

The extended duty may have been designed to prevent a landlord from avoiding liability to repair a boiler or refrigeration plant which is situated outside the physical confines of any flat, as where it is in a central basement not demised to any flat lessee.[33]

---

[29] Gas (Installation and Use) Regs 1998, SI 1998 No 2451. See Driscoll (1998) 142 Sol Jo 1014.
[30] *Sheldon v West Bromwich Corporation* (1973) 25 P&CR 360 (burst water tank).
[31] See eg *Liverpool City Council v Irwin* [1977] AC 239, HL.
[32] *Wycombe Area Health Authority v Barnett* [1982] 2 EGLR 35, CA.
[33] So reversing *Penn v Gatenex Co Ltd* [1958] 2 QB 210, CA.

Liability to repair imposed by statute

The extended obligation is, however, limited by section 11(1A)(b) of the 1985 Act:

(a) to an installation which forms part of any building in which the landlord has an estate or interest (freehold or leasehold);
(b) to an installation owned by the lessor or under his control (as no doubt where the installation is hired).

Additional limits on implied covenants

There are a number of specified additional limits on the landlord's implied covenants resulting from the 1985 Act (s 11(2)).

1. The landlord is not liable for any repairs which result from a failure by the tenant to use the premises in a tenant-like manner.
2. If the premises are destroyed or damaged by fire, tempest, flood or inevitable accident, the landlord is not obliged to rebuild or reinstate the premises.
3. The landlord is not statute-bound to keep in repair or to maintain any tenants' fixtures.

Common law limits

As with any covenant to keep in repair, the landlord is not bound to carry out work, notably to the structure and exterior of the "dwelling-house", which is too extensive to be a repair, nor wholly to improve the "dwelling-house" (see further Chapters 4 and 5). Thus, where a landlord voluntarily carried out extensive work of renovation to a house, extending its life and improving it, he was not liable to pay damages to the tenant for the cost of redecorating the interior,[34] which damages he would, as seen, have had to pay had the work fallen within the statutory covenant.

It has been held to follow from the reference in section 11(1) to "repair" that a landlord was not bound to replace undamaged metal-frame windows with wooden-frame windows so as to ameliorate severe condensation in a house built to correct design standards at the time.[35] The fact that the house was unfit for human habitation and yet in repair for the purposes of the statutory

---

[34] *McDougall* v *Easington District Council* [1989] 1 EGLR 93, CA.
[35] *Quick* v *Taff-Ely Borough Council* [1985] 2 EGLR 50, CA.

covenant exposes, as we have seen, a lacuna in the law. The courts are not disposed to change their interpretation of a covenant to keep in repair merely because Parliament has neglected to revise the statutory rent limits applying to unfit housing. It was even observed that "if the only defect in the door was that it did not perform its primary function of keeping out the rain, and the door was otherwise undamaged ... this cannot amount to a defect [for statutory purposes]".[36]

Flats: extension of duty to repair

The statutory covenant respecting the "structure and exterior" was found to be insufficiently wide in the case of short leases of flats as opposed to individual houses because it had been held that the landlord's obligations extended only to the "structure and exterior" of any individual flat.[37] So, in the case of a ground-floor, as opposed to a top-floor flat immediately under the roof, a common roof to both flats did not form part of the "structure and exterior" of the ground-floor flat and the landlord could not be compelled by the tenant of the latter to repair the roof. Similarly, exterior walls of the block containing the flat which were not the external walls of the subject flat arguably fell outside the scope of section 11.

To remedy these difficulties, the scope of section 11 was extended so as to apply to short leases of flats entered into as from January 15 1989, but not to leases entered into before that date. The landlord's statutory implied covenant to keep in repair the structure and exterior in the case of flats is to extend to "any part of the building" in which the landlord has an estate or interest (s 11(1A)(a)). Such estate could be freehold or leasehold.

A landlord's breach of statutory covenant in respect of external walls or the common roof or common parts of a whole block of flat units may be enforced by any short lessee of a flat in the block, provided, however, that the disrepair is "such as to affect the lessee's enjoyment" of his flat or common parts he is entitled to use (s 11(1B). Hence, if a leak in a common roof causes no disrepair or damage to the complainant's flat, he has no cause of action under the Act for the time being.

---

[36] *Stent* v *Monmouth District Council* [1987] 1 EGLR 59 at 64K, CA; in fact the landlord voluntarily replaced the disputed door with an improved door.
[37] *Campden Hill Towers* v *Gardner* [1977] QB 823, CA.

Landlord's defence to extended obligations

The landlord may require to gain access to carry out repairs to a part of the block not demised to the flat lessee who complains that disrepair is affecting his enjoyment of his flat (or his prescribed installations). The landlord has a statutory right to enter and inspect the premises demised to the complaining lessee (s 11(6)), which will not assist him where there are leases of flats that are not subject to section 11 of the 1985 Act. These leases may not in terms allow the landlord any right of entry or access for the purpose of carrying out repairs. To meet this type of difficulty, the landlord has a statutory defence, in any proceedings relating to failure by him to comply with his obligations. He must prove: "that he used all reasonable endeavours to obtain, but was unable to obtain" such rights "as would be adequate to enable him to carry out the works or repairs" (s 11(3A)). Thus a landlord could argue that he was unable to comply with his extended obligation because his access could not be gained as of right against a given lessee of another part of the block and that access by permission was refused.

If, however, the disrepair is located in a flat which adjoins or is immediately above or below the complaining lessee's flat, and its remedying is essential preservation work, and since the landlord has a reversionary estate in the complainant's flat, the complainant might be able to argue that any failure or refusal by the landlord to seek an access order against the neighbouring lessee under the Access to Neighbouring Land Act 1992 would prevent the landlord from avoiding liability, at least in damages, in subsequent proceedings.

Entry by landlord to repair

The landlord or any person authorised by him in writing has a statutory right to enter and inspect the premises, at reasonable times of day, to view their condition and state of repair, upon giving 24 hours' written notice (s 11(6)). This overriding privilege carries with it an implied right to enter and execute the repairs. The landlord may accordingly insist, against the tenant's will, on entering and carrying out the required repairs and commits a trespass only if the work goes outside the scope of his statutory duty.[38] However, the existence of this right of entry has had no effect

---

[38] *McDougall v Easington District Council, supra.*

on the notice principle: there is no implied duty as a result of it on a landlord to inspect the premises regularly, which he ordinarily should do in the case of an express covenant to keep in repair, since, as noted, liability normally arises as from the time damage takes place, not some later time when it is cured.

Notice rule

In contrast to the position arising under section 4 of the Defective Premises Act 1972, the landlord is not liable to carry out any of the repairs required of him by the 1985 Act, unless he has notice of the want of repair, from the tenant or a third party such as his rent collector or agent. Notice means actual knowledge, and involves the landlord having sufficient notification or information of the fact of disrepair to put him on inquiry as to the nature of the disrepair, and as to the remedial work required. Formal notice is not necessary for present purposes.[39] Knowledge by the landlord is required whether the defect is known to the tenant or not. Traditionally, the notice principle is grounded on the fact that the tenant has possession.[40] The continuing strength of this approach is such that it has recently been reaffirmed in Queensland, and without the slightest qualms.[41]

The requirement of notice is exceptional, owing to the principle that liability under a covenant to keep in repair arises, ordinarily, as soon as the defect exists which calls for remedial action.[42] Moreover, because notice is only required in respect of defects arising within the confines of the demised premises, a local authority landlord which retained the roof of the building concerned, where water pipes burst, damaging the tenant's flat, was liable as from the date of the damage arising, even though in due course the defect was put right.[43] The notice exception has been justified on the ground

---

[39] *Hussein v Mehlman* [1992] 2 EGLR 87 at 92B; Sedley J said *obiter* that a landlord who wilfully shut his eyes to damage of which a prudent property owner would have taken notice would be fixed with knowledge for the present purpose.
[40] *McCarrick v Liverpool Corporation* [1947] AC 219, HL; *O'Brien v Robinson* [1973] AC 912, HL. See also the discussion in Chapter 4 of this book.
[41] *Austin v Bonney* [1999] QdR 114.
[42] *British Telecom plc v Sun Life Assurance Society plc* [1995] 2 EGLR 44, CA.
[43] *Passley v London Borough of Wandsworth* (1996) 30 HLR 165, CA.

that it reflects common sense – a person cannot repair a defect of which he is has no knowledge.[44] It is not likely to encourage landlords to inspect the premises.[45]

III – Landlord's duty under Defective Premises Act 1972

At common law, the landlord is not generally liable in tort for personal injuries caused to a tenant or any member of his family or household by a defective and dangerous state or condition existing in any part of the demised premises,[46] as where a floor is rotten and so dangerous or where a glass panel has been defectively installed by the landlord. It was to mitigate this rule that section 4 of the Defective Premises Act 1972 was passed. The benefit of the section extends to all of the premises demised to the tenant,[47] in contrast to the position under section 11 of the Landlord and Tenant Act 1985. However, there is still nothing contrary to law in the letting of an unfit house, and until Parliament changes the current state of the law, injuries to health as opposed to personal safety will remain unremedied in private law.[48]

Scope of duty

Section 4(1) of the 1972 Act provides that where premises are let under a tenancy (widely defined by s 6(1)) which puts the landlord under an obligation to the tenant for the maintenance or repair of the premises, the landlord owes to all persons who might reasonably be expected to be affected by defects in the state of the premises a duty to take such care as is reasonable in all the circumstances to see that they are reasonably safe from personal injury (defined in s 6(1)) or from damage to their property caused by a "relevant defect". The duty is fault-based, not absolute. It only

---

[44] *Hussein* v *Mehlman* [1992] 2 EGLR 87 at 92B.
[45] *McGreal* v *Wake* [1984] 1 EGLR 42 at 43F: the rule "penalises the conscientious landlord and rewards the absentee".
[46] *Cavalier* v *Pope* [1906] AC 428. This rule is "entrenched" where not confined by statute or the common law: *Boldack* v *East Lindsay District Council* (1999) 31 HLR 41 at 49, CA.
[47] *Smith* v *Bradford Metropolitan Council* (1982) 44 P&CR 171, CA.
[48] As recognised by Brooke LJ in *Issa* v *Hackney London Borough Council* [1997] 1 All ER 999 at 1007–1008.

applies where the landlord is under an express, implied or statutory obligation to repair or keep in repair, impliedly excluding many business leases. The statutory duty arises only if there is a state of disrepair to the item concerned. A plaintiff is not able to claim under this Act on account of the fact that the premises may be unfit for human habitation.[49] In addition, the landlord may comply with his duty by rendering the item concerned safe, rather than by removing or replacing it, since he is to take reasonable care to see that the persons are reasonably safe from the injury concerned.

The duty is owed if the landlord knows of the "relevant defect" (whether as the result of notice from the tenant or otherwise). It is further expressly owed if the landlord "ought in all the circumstances to have known" of the defect (s 4(2)). If the landlord could reasonably have foreseen, because of the construction of the property, that a floor would be rotten, he must inspect the premises. If he does not, he is liable under section 4 if the defect injures any person on the premises or their property.[50] Where an obligation arose, if at all, under a previous tenancy, it was not imputed by the landlord's statutory right of entry into the current tenancy. Any injury complained of had to be related to a landlord's breach of duty under the current tenancy.[51] It is not clear whether the 1972 Act applies to latent, as opposed to patent defects. It has been said that it does not apply to latent defects, as the landlord is not a guarantor of the safety of the tenant.[52]

Extension of statutory duty

By section 4(4) of the 1972 Act, if under the tenancy the landlord has an express or implied right of entry (by statute or common law), he is treated for the purposes of this statutory liability and for no other purpose "as if he were under an obligation to the tenant for that description of maintenance or repair of the premises". The duty

---

[49] *McNerny v Lambeth London Borough Council* [1989] 1 EGLR 81, CA; [1989] Conv 216 (PF Smith).
[50] *Clarke v Taff-Ely Borough Council* (1980) 10 HLR 44.
[51] *Boldack v East Lindsay District Council* (1999) 31 HLR 41, CA (hence, injuries suffered by a plaintiff child from a paving slab leaning against the exterior of a house where works had been carried out prior to the tenancy fell outside s 4).
[52] Law Comission Report (1996), para 5.26 (in relation to unknown defects).

arises as from the time when the landlord first "by notice or otherwise can put himself in a position to exercise" such right of entry. If there is at common law an implied right of entry against a landlord of a weekly or other periodic tenant, to support an implied repairing obligation, the landlord is then to be treated as liable under the 1972 Act to the extent that the "relevant defect" has continued as from the time he ought to have entered and carried out remedial work to remove the danger caused by the defect.[53] The contrast with the notice rule applying under the 1985 Act will be apparent.

"Relevant defect"

The expression "relevant defect" is defined (s 4(3)) as a defect in the state of the premises existing on, or after, the "material time" (usually the commencement of the tenancy) and arising from, or continuing because of, an act or omission by the landlord which would constitute a failure to carry out his obligation to the tenant for the "maintenance or repair" of the premises. This definition limits the obligation of the landlord to cases of proved disrepair in the premises in breach of his covenant, express or implied, as opposed to a state of unfitness for habitation or use. The landlord is not liable for defects arising or continuing due to the tenant's failure to comply with his own express obligations under the tenancy (s 4(4)).

One effect of section 4, which cannot be contracted out of (s 6(3)), is to allow a member of the tenant's family, or the tenant,[54] to claim damages if he is injured by a defective and dangerous item, such as a glass panel or a damaged floor which the landlord was required or undertook to keep in repair, but he failed to do so, or which was dangerous when he installed it into the premises. The statutory duty is in terms in addition to any duty owed by the landlord under any other rule or enactment (s 6(2)).

If the defect complained of is situated in premises which have not been demised to the tenant, but the landlord retains control of them, as with a common staircase, and the tenant or other person is injured by a dangerous defect there, which the landlord ought to have removed or rendered harmless, the plaintiff cannot claim

---

[53] *McAuley v Bristol City Council* [1991] 2 EGLR 64, CA.
[54] *Smith v Bradford Metropolitan Council, supra.*

under the 1972 Act, for reasons explained. He should be able to claim that the landlord is subject to the duties owed under the Occupiers' Liability Act 1957.[55]

IV – Incidence of liability for repairs in particular classes of residential leases

A – *Long leases subject to modified assured tenancy rules*

As from January 15 1999, once a long residential tenancy comes to an end, it is continued as an assured weekly tenancy.[56] Such continuation is seemingly intended as an interim measure. The landlord is entitled to serve a notice terminating the continuing tenancy and proposing different terms as to repair to those of the original tenancy, which proposals may be contested by the tenant by means of a prescribed form counter-notice. Both parties have to serve their respective notices rapidly if there is a dispute, since if the tenant does not serve a counter-notice contesting the landlord's proposals within two months of service of the landlord's notice, the terms proposed in the landlord's notice take effect as the terms of the new assured tenancy. The landlord may refer any matters still in dispute to a rent assessment committee but has only two months from the service of the tenant's counter-notice in which to do so, otherwise the new terms proposed by the tenant take effect. It is provided that the rent assessment committee must consider whether the terms proposed are such as in their opinion "might reasonably be expected to be found in an assured monthly tenancy of the dwelling-house" (Sched 10 para 11(3)). In view of the fact that statute protects short residential tenants, it could be argued that continuation assured tenants should be assumed to have equally light repairing obligations, especially in view of the relative insecurity of their tenancy: thus if a landlord proposes increasing the burdens of repairing obligations on his tenant, he may have an uphill struggle persuading a rent assessement committee of the merits of his case, but there is currently no authority in point.

---

[55] See Law Commission Report (1996), para 5.27, citing from p 256 of Chapter 14 of the previous edition of this book.
[56] Local Government and Housing Act 1989, Sched 10, which is saved by Housing Act 1988, Sched 2A, para 6, from the presumption of Housing Act 1988, s 19A, to the effect that assured tenancies are to be assured shorthold tenancies.

B – Residential tenancies

If a residential tenancy is for a term of less than seven years then section 11 of the Landlord and Tenant Act 1985 will apply to the landlord. In the case both of protected or statutory and assured shorthold and assured tenancies, there are certain specific statutory rules with respect to the impact of repairing obligations.

Rent Act tenancies

1. If a contractual regulated tenancy is for a term of less than seven years,[57] so that section 11 of the 1985 Act applies to it, its terms are carried forward into any statutory tenancy arising under the Rent Act 1977 (s 3(1)). It is not possible for the parties to vary these except by agreement and except where any variation would not conflict with section 11 of the 1985 Act. The landlord has a statute-implied right of access in the case of a statutory tenancy in order to execute repairs (s 3(2)). If the tenant refuses to allow such access, this would constitute a breach of his tenancy obligations and would enable the landlord to claim possession in proceedings, subject to the overriding discretion of the court.[58]
2. Either or both parties to a regulated tenancy (ie a protected or statutory tenancy) are entitled to require the ascertainment and registration of a "fair rent" for the house or flat concerned, and such rent has a two-year life from the date of registration (s 67(3)). The landlord may apply alone for a revision of a registered rent within the two-year period only if he shows that he has carried out major improvements to the premises or rebuilt them, but not where he has simply refurbished them.[59] But if the effect of refurbishments is to raise the letting value of the house or flat, there is nothing to preclude such effect being taken into account at a regular revision of the registered rent.
3. So as not to reward the landlord for his own neglect to repair, the rent officer or rent assessment committee is required to take into account, in assessing a "fair rent", the state of repair of the

---

[57] Most cases coming before the county courts today are likely to involve statutory rather than protected tenancies: *White* v *Wareing* [1992] 1 EGLR 271 at 272H.
[58] *Empson* v *Forde* [1990] 1 EGLR 131, CA.
[59] *Rakhit* v *Carty* [1990] 2 EGLR 95, CA.

house or flat (s 70(1)(a)) as at the date of the hearing. Although the best evidence of a fair rent is taken to be supplied by assured tenancy comparables (where available),[60] the effect of the state of repair of the subject premises remains a relevant factor in the assessment process – as well as that of any comparable premises. Any effect on rent of a breach of repairing obligation by the tenant or a predecessor in title of his, whether under the current, or, seemingly, a previous tenancy, is to be left out of account (s 70(3)(a)). The tenant may show that he was unable to comply with his covenant because of the landlord's own breach. It is open to a rent assessment committee to take into account that the tenant is subject only to limited repairing obligations as opposed to the landlord's more extensive statutory duties.[61]

Assured tenancies

As from February 28 1997, when a revised assured tenancy regime came into effect,[62] it is presumed, subject to a number of exceptions, that any assured tenancy will take effect as an assured shorthold tenancy. The duration of most of these tenancies, which may be fixed-term or periodic, is not likely to be long. In these circumstances, the principal provision affecting the tenancy is likely to be section 11 of the Landlord and Tenant Act 1985. This provision cannot be altered or contracted out by the parties save with the sanction of the county court.[63] Where a fixed-term assured shorthold tenancy has been granted, the original rent level, unless there are rent reviews, is that agreed by the parties. As from the end of the tenancy, if the assured shorthold tenant stays in possession as a periodic tenant, the landlord may, following a prescribed form notice,[64] require the rent to be revised under section 13 of the 1988

---

[60] See *Curtis v London Rent Assessment Committee* [1997] 4 All ER 842, CA; subject to the limits imposed by Rent Acts (Maximum Fair Rent) Order 1999, SI 1999 No 6.
[61] *Firstcross Ltd v Teasdale* [1983] 1 EGLR 87.
[62] From the commencement of Housing Act 1996, adding s 19A to Housing Act 1988.
[63] Sections 5 and 6 of the 1988 Act only therefore seem to apply where the duration of the tenancy is for seven years or more. They enable the terms of the tenancy as to repairs to be revised one year after the initial fixed term has expired.
[64] Under Assured Tenancies (Forms) Regs 1988, SI 1988 No 2203.

Act. The rent assessment committee, charged with this task, is bound to disregard any reduction in the value of the dwelling-house attributable to a failure by the tenant, in our context, to comply with his repairing obligations (s 14(2)(c)). This disregard applies only to the current tenant, so that a rent assessment committee was not entitled to take into account, in fixing a revised rent, breaches of covenant committed by his predecessor in title, who had held a statutory tenancy.[65] In the case of any assured shorthold tenancy for less than seven years, owing to the restricted nature of the tenant's obligations, any such reduction, deliberate damage apart, is not likely to be great.[66]

Secure tenants – General

Secure tenants within Part IV of the Housing Act 1985 are subject to the same general rules as apply to any private sector residential tenants. Thus where the tenancy is periodic, the repairing obligations of the parties will be subject to section 11 of the Landlord and Tenant Act 1985. There are no provisions enabling the terms of secure tenancies to be revised as exist in the case of private sector residential tenancies.

Secure tenants' repair scheme

A repairs scheme has been promulgated in regulations, applying to those secure tenants who hold from local authorities, in force as from April 1 1994.[67] The regulations are confined to secure tenants or introductory tenants who hold from local authority landlords, and so do not apply to housing association tenants, for example, except for those with less than one hundred secure tenants at the date of the tenant's request for repairs (regs, para 3(2)). Local authorities may need to maintain lists of approved contractors (as indeed envisaged

---

[65] *N&D (London) Ltd v Gadsdon* [1992] 1 EGLR 112, so that the rent proposed by the landlord was reduced to £5 a month.
[66] There is provision (s 14(2)(b)) so as to disregard from the effect on rent voluntary improvements carried out by the tenant who at the time undertook them (so including eg not just the current assured shorthold tenant but any predecessor of his under that, or any previous tenancy).
[67] Secure Tenants of Local Authorities (Right to Repair) Regulations 1994, SI 1994 No 133, as amended.

by section 96(3)(a) of the 1993 Act but not, seemingly, by the regulations) so as to be able to comply with the scheme.

The procedure relates only to prescribed qualifying repairs within section 96 of the Housing Act 1985. A detailed and somewhat cumbersome procedure is laid down. The secure tenant applies to the landlord (there is no prescribed form) for the repair to be carried out. The landlord, subject to a right of inspection "forthwith" of the house or flat to see if the repair qualfies, must issue a repair notice to a contractor with a copy of it to the tenant, who receives an explanation of the regulations (postal service being allowed by para 9). If, however, the landlord is satisfied that the work falls outside the scheme, he must notify the tenant of this and give the tenant an explanation as well as an explanation of the regulations (regs, para 5).

A repair notice must specify prescribed particulars, such as the nature of the repair and the name, address and telephone number of the contractor, as well as access arrangements to the premises (para 5). The importance of the latter arrangements is that if the tenant has, when informing the landlord of the need for the repair, failed to provide the landlord with details of access arrangements for a contractor, or fails to allow access for inspection by the landlord or for the repairs to be carried out, his entitlement to have that particular repair carried out within the scheme, and to compensation for default, cease to apply (para 3(3)).

The tenant has the right – seemingly at any time before the work is carried out – to inform the landlord that "he no longer wants the repairs carried out" (para 3(3)). The exercise of this right denies the tenant the benefit of the regulations but the contractual position as between the landlord and a contractor is not mentioned.

The repair must fall within Schedule 1 of the regulations and, in addition, cost no more than £250 to carry out, this figure being dependent on the opinion of the landlord (para 4). Schedule 1 lists small-scale, but significant matters of repair or maintenance, such as total loss of water supply, blocked flue to open fire or boiler, blocked or leaking foul drain, blocked sink, bath or basin, leaking roof, an insecure external window, door or lock, loose or detached bannister or hand rail or a rotten timber flooring or stair tread. Against each item appears a precribed period of working days, ranging from one day (as with total loss of water supply) to seven days (for leaking roofs).

The repair notice must specify the last day of the "first prescribed period" (para 5(1)). Where there is no landlord's inspection, this is the period, ordinarily running from the first working day after issue of the repair notice, during which the repair is to be executed (para 2).

If the contractor specified in the notice fails to remedy the defect within this prescribed period, the tenant may notify the landlord that he requires the work to be carried out by another contrator, whereupon the landlord, where this is reasonably practicable, is to issue a second repair notice to the contractor, following the same procedure as that applicable to the first such notice (para 6). If that contractor fails to carry out the work within the second prescribed period (which is similar in duration to the first such period) the tenant is entitled to be paid a modest compensation by the landlord (para 7(1)). But, and this may amount to a significant limit, the landlord is able to set off against this sum any moneys owed to him by the tenant (para 7(3)).

The specified sum of compensation is the lesser of £50 and the sum ascertained by a formula designed to penalise the landlord for each continuing day of delay in execution of the work by the contractor (para 7(2)). The payment of the compensation would presumably not enable the landlord to avoid liability for damages for consequential losses flowing from the second contractor's breach, even allowing for the tenant's duty to mitigate the loss by himself procuring the execution of the repairs, if his remedy under the regulations fails.

The prescribed periods are suspended if there are circumstances of an exceptional nature beyond the control of the landlord or the contractor which prevent the repair being carried out (para 8). The ascertainment of the duration of any such period may give rise to disputes both between tenants and landlords and between landlords and contractors, as where there are supply difficulties or strikes.

A number of uncertainties exist within these regulations. Thus, it is for the landlord to satisfy itself that a repair is needed (the criterion is subjective and not objective). The landlord is to carry out any inspection "forthwith", a requirement which cannot be taken to require literally an instant compliance. If the tenant (and so not, it seems, a third party) "informs" the landlord that repairs are not required, the regulations do not apply, but written information is not required and yet oral information may be difficult to rely on. The question of whether a repair will not cost more than £250 to remedy is simply a matter for the landlords "opinion", which does not necessarily have to be reasonably formed.

C – Liability of parties under extended leases

Extended lease under Leasehold Reform Act 1967

Where the long lessee of a house obtains an extended lease under the 1967 Act, the repairing and any service charges terms of the 50-year extended lease are in principle the same as under the expiring long lease, fixed as at the tenant's notice of desire to extend. These terms may be altered by agreement. Both parties may require the exclusion or modification of any term in the old lease or any agreement collateral to it which it is unreasonable to include, unchanged, in the extended lease, in view of the date of the original tenancy and changes since then which have affected its provisions (s 15(7)), disputes being for the county court, as opposed to the Lands Tribunal, to settle (s 20)). In contrast to the position with business tenancy renewals, a landlord could presumably ask for the substitution of a variable for a fixed service charges clause under this provision.

Lease-back under long leasehold enfranchisement

Where a nominee purchaser has acquired the freehold of a block of long leasehold flats under recent statutory procedures, rather than by voluntary agreement, the former freeholder is entitled to require a long lease-back, for 999 years, of any flats held by secure tenants (or let to non-secure tenants by a housing association) and he may require such a lease-back of flats occupied by him as resident landlord.[68] The nominee purchaser, as new lessor, is subjected to a statutory covenant "to keep in repair the structure and exterior of the demised premises and of the specified premises (including drains, gutters and external pipes) and to make good any defect affecting that structure".[69] Although the former freeholder in his capacity as long lessee is bound to bear a reasonable part of the cost of compliance with the new freeholder's obligation, it cannot be avoided by a term in the new long lease and it appears to extend far beyond mere repairs, and might well entail a guarantee that if part or the whole of the structure of the premises concerned collapses or is otherwise defective, for any cause, this will be made

---

[68] Leasehold Reform, Housing and Urban Development Act 1993, s 36 and Sched 9.
[69] 1993 Act, Sched 9 Part IV para 14(1)(a)).

good by the nominee purchaser who will then have to pass on such a cost to the acquiring tenants.

V – Effect of disrepair on rent and terms of new business tenancies

Where business premises are held by a tenant within Part II of the Landlord and Tenant Act 1954, who has applied for a new tenancy, questions arise as to the effect of any serious disrepair on the rent payable for the new tenancy under section 34 of the Act and as to the variation of the service charges obligations of lessees under section 35.

A – Rent payable

The general approach of the courts is to take into account serious dilapidations in reduction of both the interim and the ultimately-payable market rent where these are the fault of the landlord.[70] If the disrepair is serious, the court may award a "differential rent" to be paid under the new tenancy, so that the full increase in rent is put off until such time as the landlord remedies any condition of disrepair for which he was responsible under the contractual tenancy. Similarly, the court may assume that since a market rent is to be payable under section 34 of the 1954 Act, the tenant is not entitled to plead his own breaches of covenant to repair under the contractual tenancy in reduction of his rent – even though the effect of this may well be to deprive the landlord of any damages at this stage since his reversion is deemed to have suffered no loss.[71]

B – Variation of terms

The policy of the 1954 Act, according to the House of Lords, is that if the parties cannot agree as to revised terms, the court has only a limited discretion under section 35 to compel a variation of any repairing and service charge payment obligations upon an unwilling lessee, at least where the court considers that he has not been offered sufficient compensation, as in the basis of calculation of rent.[72] Accordingly, a landlord failed to persuade the House of

---

[70] *Fawke* v *Viscount Chelsea* [1979] 1 EGLR 89, CA.
[71] *Family Management* v *Gray* [1980] 1 EGLR 46, CA.
[72] *O'May* v *City of London Real Property Co Ltd* [1983] 2 AC 726.

Lords that it should order an unwilling tenant, who was not offered sufficiently generous compensation by the landlord, to accept that, in the new tenancy, the tenant should undertake the liability to pay in variable service charges clauses for repairing and maintenance work which was the liability of the landlord. Although the Law Commission, when considering this issue, noted complaints of landlords' representative groups about the fact that this principle placed a heavy onus of proof on landlords in favour of change, they did not recommend any amending legislation owing to the "necessary combination of stability and flexibility" in the current law.[73]

VI – Control of unfit housing by local authorities

Local authorities have statutory powers to deal with unfit dwelling-houses and flats, in particular under Part VI of the Housing Act 1985. The main principles applying are noted. The relevant local authority, which is required at least once a year to consider housing conditions in their district and to determine what action to take (s 605) has a choice as to which of three courses of action to adopt. They may serve a repair notice, or notices relating to a house in serious disrepair, or make a closing or demolition order.

The authority is bound, in making up its mind as to which option to take, to have regard to such guidance as the Secretary of State publishes from time to time.[74] If an authority's decision is challenged, the court asks the narrow question whether the decision-making body has reached an unreasonable decision either by taking into account irrelevant factors or omitting to take into account relevant factors, having regard in a broad sense to the Ministerial guidance. The weighing up of socio-economic factors, for example, with the cost of the work – both are said by the guidance to be of importance – is primarily a matter, subject to the reasonableness test, for the authority.[75] The imprecise nature of some of the factors idenitifed in this guidance vindicates the approach of the High Court. If a local authority intends to serve either a repair notice or a serious disrepair notice, it must serve a notice that it is minded to take action, specifying what enforcement action the

---

[73] Law Com No 208, Business Tenancies (1992), para 3.33.
[74] Housing Act 1985, s 189(1) and 604(1A).
[75] R v *London Borough of Southwark, ex parte Cordwell* (1993) 26 HLR 107.

authority is minded to take, with reasons. The person served with such a notice has 14 days to make representations to the authority, which it is bound to consider.[76]

A – Repair notices

A local authority may serve a repair notice on the owner or other person having control of the offending house or flat. The premises must be unfit for human habitation in the statutory sense. In the case of a block of flats, an authority may serve a repair notice where an individual flat is unfit because of a defective condition in part of the building outside the flat (s 189(1A)) as where a common roof or foundations are unfit. The procedure governing repair notices is the same as that applicable for notices as to disrepair not amounting to unfitness.

The statutory definition of unfitness is crucial: if the premises are not unfit then the repair notice procedure cannot be used and recourse must be had to other remedies, such as related notices or the statutory nuisance procedure. By section 604 of the Housing Act 1985, there is a presumption that the dwelling-house concerned is for human habitation unless, in the opinion of the local authority, it fails to meet one or more of a list of requirements.

The statutory list refers to: (i) structural stability; (ii) freedom from serious disrepair; (iii) freedom from dampness prejudicial to the health of occupiers; (iv) an adequate piped supply of wholesale water; (v) satisfactory facilties in the house for the preparation and cooking of food, including a sink with a supply of hot and cold water; (vi) a suitably-located WC; (vii) a suitably-located fixed bath or shower and hand-basins, all for the use of occupiers, and each of which installations is supplied with hot and cold water; and (viii) an effective system for the drainage of foul waste and surface water.

Premises may be unfit for habitation without being in disrepair. If a tenant, in the ordinary use of the premises, is caused personal injury by a defect, then the premises are presumed to be unfit for human habitation.[77] In addition, while a house which suffers from severe dampness or condensation may not be out of repair, it would be unfit for statutory purposes, so enabling the authority, if it thought fit, to serve a repair notice on the owner. The tenant

---

[76] Housing Construction Grants and Repairs Act 1996, s 86.
[77] *Summers v Salford Corporation* [1943] AC 283, HL.

cannot resort either to contractual or tortious remedies in such a case, for reasons explained: and the relevant private law statute has been recognised as a dead letter.[78]

A repair notice must comply with certain formalities, notably, that it be served on the person "having control of the house" who is then required to execute the works specified in the notice within a reasonable time. This period must be not less than 21 days and it must be specified in the notice (s 189(2)). A copy of the notice may be served on any person with an interest in the house, such as a mortgagee.

Meaning of "person having control"

The statutory definition of the "person having control" is, by section 207, the person who receives the rack-rent (as opposed to a ground or nominal rent because the sum must be at least two-thirds of the net annual value of the premises) whether on his own account or as agent or trustee for another person. The notice is to be served on the person who may ultimately have to pay for the cost of the works.[79] It is he who ought in justice to be responsible to discharge the liabilities to which the premises in their condition have given rise. A repair notice served on the freeholder of a block of residential flats, which were held by long lessees paying ground rents, was hence invalid. Notice could validly have been served on the long lessees. The specified works may be either repair or improvement, so cutting accross common law distinctions, though presumably the notice will distinguish between the two where relevant.[80]

Notices – Further

A repair notice will be invalid if it fails to contain sufficient information to enable the owner to have the work costed by a reasonably competent builder, but this requirement does not require the notice to specify the particular works, provided it lists the general nature of the works required.[81] The notice must state

---

[78] *Issa* v *Hackney London Borough Council* [1997] 1 All ER 999, CA
[79] *Pollway Nominees Ltd* v *Croydon London Borough Council* [1987] AC 79 at 92D (Lord Bridge).
[80] See *R* v *Forest of Dean District Council* The Times, November 9 1989.
[81] *Church of Our Lady of Hal* v *Camden London Borough Council* [1980] 2 EGLR 32, CA.

that in the opinion of the authority, the works will render the "dwelling-house" fit for human habitation, once completed. A repair notice becomes operative once 21 days from the date of its service expire and it becomes final and conclusive as to matters which could have been raised in any appeal against it within the 21-day period (s 189(4)). An invalid notice cannot be cured by this rule. The 1985 Act provides a detailed procedure for an "owner" to appeal against a repair notice: if an appeal is made, the notice is suspended until the appeal outcome is known. If an appeal fails, the authority has default powers to carry out works, if there is no agreement with the owner.[82]

B – Serious disrepair notices

With a view to preventing landlords of dilapidated residential premises from allowing them to fall into such a bad state and condition that no compliance with any repair notice under section 189 will be possible, leaving closing or demolition as the only possibility,[83] a local authority may serve a repair notice where it is satisfied, after a representation by an occupying tenant (or in the case of a house in multiple occupation, by occupiers whether as tenants or licensees) that the "dwelling-house" is in such a state of disrepair that, although it is not unfit in the statutory sense, its condition interferes materially with the personal comfort of an occupying tenant or occupying licensee (s 190(1)(b)).

A local authority is also empowered to serve a repair notice where it is satisfied that a house, flat or house in multiple occupation is in such a state of disrepair that, although not unfit for habitation, substantial repairs are required to bring it up to a reasonable standard, having regard to its age, character and locality (s 190(1)(a)). Both the procedure applicable to notices under section 190 and the persons on whom it must be served is largely the same as for repair notices under section 189. Thus a statutory tenant, who has no proprietary interest in premises, was not a proper recipient of a section 190 notice.[84]

---

[82] Housing Act 1985, ss 193 and 194 and Sched 10.
[83] *Kenny* v *Kingston-upon-Thames Royal London Borough* [1985] 1 EGLR 26, CA.
[84] *White* v *Barnet London Borough Council* [1989] 2 EGLR 31, CA.

Reform

The government have claimed that in practice local authorities do not make much use of their powers as to fitness for human habitation. The present repair notices regime, based as it is on general standards of fitness for human habitation, is likely, depending on the results of consultation in progress as this book went to press, to be replaced with dwelling-by-dwelling health and safety rating scales. Thus it would be possible, it is asserted, to base ratings on hazards actually faced by occupants. The intervention of local authorities could then be tailored to the severity of hazards within individual dwellings.[85] While the Law Commission reform package (which as we saw would extend the landlord's repairing obligations so as to cover housing unfitness) might be a "useful supplement" to these latest government reform ideas, it now seems that the government would only promote these particular reforms with "some modification" to take into account its own principles.[86] It may be that the Treasury prefers to dilute reform, moving from general principles to supposedly individually tailored models, whose application would be partially subjective, leaving some housing unfit but not dangerous to health, for example, for fear of the financial implications to local authorities and perhaps to owners of the Law Commission package being implemented as originally conceived.

VII – Control of statutory nuisances

A – Introduction

The rules concerning control of statutory nuisances are in the Environmental Protection Act 1990. The main principles applying are noted. The procedure for combatting serious property disrepair is useful as imposing a criminal penalty on the person responsible for continuing the statutory nuisance. Any costs of the prosecution are normally borne by the authority acting against the owner, rather than by any lessee. While the Act is not a self-contained code,[87] it does not allow any private individual, even if his health has been injured by a statutory nuisance, to bring a civil claim

---

[85] Housing Green Paper (2000), para 5.27–5.28.
[86] *Ibid*, para 5.29.
[87] See *Carr* v *Hackney London Borough Council* (1995) 93 LGR 606.

outside the 1990 Act against the person responsible for it.[88] An attempt by a tenant to persuade the court to imply into his tenancy a term which he said would abate the nuisance concerned, failed, as the court only implies terms on the ground of essential need, not expediency or to fill gaps.[89]

B – Duties of authority

A local authority (as defined in s 79(8)) may only act where there is a "statutory nuisance". This arises where "any premises [are] in such a state as to be prejudicial to health or a nuisance" (s 79(1)(a)). Sometimes, the authority will act following a complaint to them by a person living in their area. It thereafter becomes their duty to take such steps as are reasonably practicable to investigate the complaint (s 79(1)). Equally, local authorities are subject to a statutory duty to cause their areas to be inspected from time to time, so as to detect any statutory nuisances.

Enforcement proceedings

If the constituent elements of a statutory nuisance exist, or are likely to occur or recur in their area, local authorities are bound to require the person responsible for the nuisance, or the owner of the premises if different, to abate the nuisance (s 80(1)).[90] The authority proceed to serve an abatement notice (s 80(2)) on, notably, the person responsible for the statutory nuisance, or if the nuisance "arises from any defect of a structural character", on the owner of the premises. If neither of the above persons can be found, the notice is to be served on the "owner or occupier" of the premises.

Contents of abatement notice

There is no prescribed form for an abatement notice, but it should be in writing. Clarity in an abatement notice is important.[91] There is

---

[88] *Issa* v *Hackney London Borough Council* [1997] 1 All ER 999, CA.
[89] *Habinteg Housing Association* v *James* (1995) 27 HLR 299, CA.
[90] Unless the authority resorts to its power to itself abate the nuisance under s 81–81B of the Act.
[91] *Camden London Borough Council* v *London Underground Ltd* [2000] Env LR 369; also *Stanley* v *Ealing London Borough Council* (2000) 32 HLR 745.

no general duty to consult the perpetrator of the nuisance before serving an abatement notice,[92] although the authority has a discretion to do so. At least one or more of the following things must be aimed at by the notice (s 80(2)):[93] (1) abating the nuisance; (2) prohibiting or restricting its occurrence or recurrence; and (3) executing of such works or taking such other steps as may be necessary for these purposes. In all cases, the local authority can, if it wishes, leave the choice of means of abatement to the perpetrator of the nuisance. If means of abatement are required by the local authority, these must be specified in the abatement notice.[94] Magistrates may amend a defective notice,[95] which averts the need to quash it.

C – Preconditions of statutory nuisance

There can be no statutory nuisance without "premises" within the Act, which are prejudicial to health or a nuisance.

Meaning of "premises"

Apart from expressly including land (s 79(7)), the 1990 Act does not define the word "premises". The term "premises" is wide and would presumably include any permanent or even temporary structure on land. In the case of flats, unless the state of the whole block falls within the definition of statutory nuisance, "premises" is thought to mean an individual flat unit so that the owner of the whole block, as opposed to individual long lessees, could not be prosecuted because one or more individual flats constituted a statutory nuisance.

---

[91] cont.
However, it appears that an abatement notice may be read with a covering letter.
[92] *R v Falmouth and Truro Port Health Authority, ex parte South West Water Services Ltd* [2000] 3 All ER 307, CA.
[93] A managing agent is a proper recipient of a s 80(2) notice: *Camden London Borough Council v Gunby* [1999] 3 EGLR 13.
[94] *R v Falmouth and Truro Port Health Authority, ex parte South West Water Services Ltd* [2000] 3 All ER 307, CA (where the position was reviewed).
[95] Statutory Nuisance (Appeals) Regulations 1990, SI 1990 No 2276, reg 2(5).

"Prejudicial to health or a nuisance"

These words in section 79(1) were considered in pre-1990 Act decisions concerned with identical previous words, and which held that dilapidated premises may be "prejudicial to health" if there is a threat of disease, vermin or the like,[96] as well as lack of proper sound insulation in a roof void.[97] Any condition complained of must amount to a public or private nuisance.[98] Accidental physical injury (as caused or risked by a steep internal staircase) lies outside the statutory nuisance provisions as a matter of authority.[99] Nor is it enough to show that the premises are seriously dilapidated or even unfit for human habitation,[100] unless the premises are in a state to be prejudicial to health. If that is so, as with a flat which suffered from damp and fungal growth due to inadequate ventilation, then a statutory nuisance *prima facie* exists,[101] even though the premises might not be out of repair.

Premises not in prejudicial state

Where the state of the premises, however dilapidated, or unfit for habitation or use, is not such at the date of the hearing[102] as to be prejudicial to health, then they must consitute a nuisance in the public or private sense or there is no statutory nuisance. Therefore the condition of the premises, to amount to a "nuisance" within this part of the definition, must be such as to cause damage or injury to adjoining or neighbouring premises.

Defences to liability

The person otherwise responsible for continuing a statutory nuisance has a defence to liability if he shows that the injury is the

---

[96] *Coventry City Council v Cartwright* [1975] 2 All ER 99.
[97] As in *Network Housing Association Ltd v Westminster City Council* (1994) 93 LGR 280.
[98] *NCB v Thorne* [1976] 1 WLR 543; *Wivenhoe Port Ltd v Colchester Borough Council* [1985] JPL 396.
[99] *R v Bristol City Council, ex parte Everett* [1999] 3 PLR 14, CA.
[100] *Salford City Council v McNally* [1975] 2 All ER 860, HL.
[101] As in *Greater London Council v Tower Hamlets London Borough Council* (1983) 15 HLR 57.
[102] *Carr v Hackney London Borough Council* (1995) 93 LGR 606.

result of the tenant's own fault, as where a landlord proved that his tenant was making an excessive use of a heating system, which was not itself faulty and which was intended only for background use.[103] It is a defence that the person otherwise responsible for the statutory nuisance has made reasonable efforts to abate it, as by reasonably offering suitable alternative accommodation to the complainant.[104]

D – Appeals against notice and enforcement

The person served with an abatement notice has 21 days from the date of its service on him in which to appeal against it (s 80(3)). The effect of an appeal is to delay any enforcement proceedings pending the final determination of the appeal. It is a criminal offence for the person in question "without reasonable excuse" to contravene or to fail to comply with "any requirement or prohibition imposed by the notice". It now appears that the prosecution must show absence of reasonable excuse, to criminal standards of proof (ie beyond reasonable doubt) if this defence is pleaded.[105] The courts have refused to supply a list of which matters might or might not amount to a reasonable excuse: the issue is one of fact in each case.[106] Where an owner procured the emptying of the offending premises, there remained a statutory nuisance, as the premises could be reoccupied in the future.[107] The statutory defence might apply if the owner made the premises impossible ever to be occupied in the future, as by destroying them.[108]

Because it is provided (s 81(1)) that where more than one person is responsible for a statutory nuisance, an abatement notice may be served on each such person, the number of notices and of

---

[103] *Dover District Council v Farrar* (1982) 2 HLR 32; see further Luba and Knafler p 182ff; also *Pike v Sefton Metropolitan Borough Council* [2000] EGLR Dig 272 (condensation from tenant's reluctance to pay for adequate heating).
[104] Under s 80(7) of the 1990 Act, it is also a defence in the case of premises used for industrial, trade or business purposes that the person has used "the best available means to protect or counteract" the nuisance.
[105] *Polychronakis v Richards & Jerron Ltd* [1998] Env LR 347.
[106] *Butuyuyu v Hammersmith and Fulham London Borough Council* (1997) 29 HLR 584, CA.
[107] *Lambeth London Borough Council v Stubbs* (1980) 78 LGR 650.
[108] See *Coventry City Council v Doyle* [1981] 1 WLR 1325.

convictions may be extended, as where a freeholder and lessee both continue a statutory nuisance. In addition, in extension of criminal liability, an owner of premises which constitute a statutory nuisance may personally be held liable for his own failure to abate the nuisance, regardless of the liability of any other person, such as an occupying lessee, to abate it.[109]

The penalty for infraction is a fine, with a continuing penalty for each day of continuance after conviction, but, in keeping with the relative leniency of the Act in respect of industrial, trade or business premises, the latter "ratchet" effect is excluded in relation to convictions for a statutory nuisance committed on such premises (s 80(5) and (6)), but the general penalty imposed in such cases is heavier than where the premises are, say, domestic. The complainant, if injured by a proved statutory nuisance, is, despite ambiguities in section 82 not present in the earlier legislation, still entitled to be paid compensation.[110]

---

[109] *Clayton* v *Sale Urban District Council* [1926] 1 KB 415.
[110] *Botross* v *Hammersmith and Fulham London Borough Council* (1994) 93 LGR 268.

Chapter 11

Party walls and dangerous structures

I – Background to statutory rules

A – *New statutory framework*

This Chapter discusses the main rights and liabilities of owners of party walls and outlines the control by local authorities of dangerous structures.

In the case of party walls, separate legislative rules formerly applied to Inner London and to the rest of England and Wales. The rules governing Inner London were regulated by the London Building Acts 1930–1982, of which the principal Act was the London Building Acts (Amendment) Act 1939. The 1939 Act provided for a distinct code to regulate the mutual rights and remedies of party wall owners in Inner London, with a specific disputes resolution procedure. The mutual rights and remedies of party wall owners outside Inner London were regulated by the common law as modified by section 38 of the Law of Property Act 1925. However, as from July 1 1997,[1] the whole of England and Wales was subjected by the Party Wall etc Act 1996 to a codified system enabling work to be done to party walls, party fence walls and party structures. It provides for a specific disputes regulation scheme. The 1996 Act is based on the 1939 Act system which has been improved in a number of points of detail. It was said in relation to the 1939 Act that it worked well, and that although there is provision for notices and dispute resolution, it was expected that the exercise of statutory rights would be agreed.[2]

---

[1] Party Wall etc Act 1996 (Commencement) Order, SI 1997 No 670. See *Party Walls, The New Law*, Bickford-Smith and Sydenham (1997); also Anstey, *Party Walls*, 4th edn (1996).
[2] *Chartered Society of Physiotherapy* v *Simmonds Church Smiles* [1995] 1 EGLR 155 at 157C.

B – General aspects of the statutory regime

The 1996 Act sets out specific rights in relation to work on party walls and the like for building owners. These are exclusive of the common law,[3] but the Act does not authorise any interference with an easement of light or other easements in relation to a party wall (s 9). Thus, if a right to support of an adjoining owner is in fact interfered with by the works of a building owner, or he commits a nuisance in the course of his works, the adjoining owner may bring an action for an injunction or, if adequate compensation, damages.[4]

When deciding if a wall is a party wall, it was said in relation to a predecessor to the 1996 Act that it was not necessary, when dealing with disputes under the statute, as it might have been at common law,[5] to consider questions of ownership. The court considers the physical condition, position and user of the wall.[6] It is curious to note that a "party fence wall" (s 20 of the 1996 Act) is an expression which refers to a wall "which stands on the lands of different owners and is used or constructed to be used for separating such adjoining lands" – a structure which would have been treated at common law as a party wall of the fourth type within the common law classification.

No implied duty to execute repairs

At common law, an owner of a party wall was not bound to execute any repairs to his part of a party wall necessary to ensure the enjoyment by the adjoining owner of any implied easements of support and user.[7] But should an owner now so neglect his party wall, where it is built over the boundary between the two sets of premises, that the adjoining owner needs to exercise his statutory right (under s 2(2)(b) of the 1996 Act) to repair or rebuild it, he would be at risk of having to bear a portion of the expenses (s 11(5)).

---

[3] *Standard Bank of British South America (Africa)* v *Stokes* (1878) 9 ChD 68 at 73 (Jessel MR), referring to the Metropolitan Buildings Act 1855.
[4] See *Benzie* v *Happy Eater Ltd* [1990] EGCS 76 (where there was an unsuccessful claim of interference with natural rights of support).
[5] See the four-fold classification in *Watson* v *Gray* (1880) 14 ChD 192, in relation to the rules then applicable outside Inner London.
[6] *Knight* v *Purcell* (1879) 11 ChD 412 at 415 (Kay J).
[7] *Jones* v *Pritchard* [1908] 1 Ch 630; also *Rhone* v *Stephens (Executrix)* [1994] 2 EGLR 181, HL.

Withdrawal of support and repairs

The 1996 Act, with its notice scheme, has overtaken the common law principle that a neighbouring owner could with total impunity demolish his party wall.[8] At common law, it had been held that if a wall fell down due to want of repair, not withdrawal of support, only if the defendant could reasonably have foreseen damage being occasioned to the plaintiff's premises due to the collapse will he be liable to the plaintiff – he must be shown to have been at fault.[9] Thus, an owner was not bound to prevent his part of the wall from becoming destroyed by the ordinary ageing process. It appears, in the absence of any authority, that these principles would apply under the 1996 Act to the extent only that if a party wall owner wishes to prevent a collapse, it is for him to invoke the procedures of the Act. If a building owner incurs expenses in the process, for example, of repairing or rebuilding a dilapidated party wall, pursuant to section 2(2)(b) of the 1996 Act, he has a right of contribution from the adjoining owner under section 11(5). Liability to pay is apportioned in the same way as under section 11(4), which we next note. Where work involves the underpinning, thickening or raising of a party wall structure, party fence wall or external wall built against either, then the cost is apportioned by section 11(4) between the two owners in such proportion as has regard to two factors:

(a) the use which the owners respectively make or may make of the structure or wall concerned; and
(b) responsibility for the defect or want of repair concerned, if more than one owner makes use of the structure or wall concerned.

II – Statutory regulation of rights of party wall and party fence wall owners

A – *Purpose of Act*

A principal object of the 1996 Act is to provide for a self-contained list of rights for building owners in relation to party walls and related structures, combined with a *sui generis* scheme for dispute resolution. In return, the Act makes the exercise of the various, and

---

[8] See *Smith* v *Thackerah* (1866) LR 1 CP 564.
[9] *Sack* v *Jones* [1925] Ch 235.

extensive, rights of a building owner dependent on notice procedures. Sometimes he has a right to contribution to the cost of the work, at least where the adjoining owner benefits from it or had been at fault, so causing it. Within the Act appear to be party walls in the sense most often understood under the former common law, viz, those built crossing the boundary of two adjacent premises. Also included are party walls or structures wholly built on one owner's land separating it from adjacent premises, and structures which are dividing walls between houses or horizontal dividing floors between flats, wherever the dividing line between the premises precisely falls. The 1996 Act provides a specific disputes resolution procedure, triggered by notices and counter-notices and leading to the appointment of surveyors who make an award. It is thereby hoped to avoid litigation as an instrument of first resort.

B – Statutory definitions

The expression "party wall" is defined by section 20 of the 1996 Act as:

> (a) a wall which forms part of a building and stands on lands of different owners to a greater extent than the projection of any artificially formed support on which the wall rests; and
> (b) so much of a wall not being a wall referred to in paragraph (a) above as separates buildings belonging to different owners.

The rules of the 1996 Act apply also to "party fence walls". These are defined in section 20 as follows:

> A wall (not being part of a building) which stands on lands of different owners and is used or constructed to be used for separating such adjoining lands, but does not include a wall constructed on the land of one owner the artificially formed support of which projects into the land of another owner.

"Party structure", which is also within the scope of the Act, means (s 20) "a party wall and also a floor partition or other structure separating buildings or parts of buildings approached solely by separate staircases or separate entrances."

The Act gives various rights to a "building owner" and refers to an "adjoining owner". The expression "owner" is defined in section 20, in a revised form, as follows. It in terms includes:

> (a) a person in receipt of, or entitled to receive, the whole or part of the rents or profits of land;

(b) a person in possession of land, otherwise than as a mortgagee or a tenant from year to year or for a lesser term or as tenant at will;
(c) a purchaser of an interest in land under a contract of purchase or an agreement for a lease, otherwise than under an agreement for a tenancy from year to year or for a lesser term.

This definition is not exhaustive, but it confirms that a person entitled to a lease for five years is as much entitled to a building owners' notice as would be a freeholder.[10] A long lessee who sublet the land but who received rents was entitled to the service of statutory notices and, where relevant, was bound to contribute towards the cost of works allowed under the Act.[11]

The definition which section 20 of the 1996 Act replaces, that in section 5 of the London Building Act 1930, referred to an "owner" as including "every person in possession ... or in the occupation of land"; whereas the words in section 20(1)(c) did not appear in the 1930 Act. These changes are not thought to make any difference to the result of a High Court ruling that where a "building owner" consisted of two joint tenants, both must join in the service of a statutory notice to enjoy rights under the Act.[12] It is impossible to regard one joint tenant alone as being in possession of his property. The concept of one joint owner being able to deal with the property without the other being party to the transaction was thus alien to English law. The court placed a restrictive construction on the now redundant term "occupation". Its disappearance amounts to statutory affirmation of the strict view taken in this case.

The statutory expression "building owner" means (s 20) "an owner of land who is desirous of exercising rights under this Act". "Adjoining owner" and "adjoining occupier" mean "any owner and any occupier of land, buildings, storeys or rooms adjoining those of the building owner". A person wishing to exercise rights under the 1996 Act may, thanks to these wide definitions, have more than his immediate adjacent neighbour on the other side of a party wall to consider as proper recipients of notices.

---

[10] See *Fillingham v Wood* [1891] 1 Ch 51.
[11] *Hunt v Harris* (1865) 19 CB NS 13.
[12] *Lehmann v Herman* [1993] 1 EGLR 172. Hence, a notice served by one joint tenant alone was invalid and he could not start the relevant works.

Effect of Act on existing property rights

Although party wall legislation supercedes the common law in relation to any matter falling within its scope, giving a comprehensive code,[13] by section 9, the 1996 Act does not authorise any interference with an easement of light or other easements in or relating to a party wall (such as support). Nor does it prejudicially affect the right of any person to preserve any right or thing in or in connection with a party wall where a party wall is rebuilt. Thus, an award which purported to eliminate light enjoyed as of right in an unlimited amount through the lessees' windows was held to be invalid.[14] This legislation does not affect the legal title to a party wall or party structure.[15] It limits the building rights of owners for the general benefit of the public.[16] Whether a wall is a party wall or not depends on its user. It may be a party wall for part of its height and above that simply an external wall – that latter part being wholly the property of the relevant owner.[17] Hence, a building owner could not legitimately claim that a neighbouring owner should contribute to the cost of raising a wall above the level where it ceased to be a party wall.[18]

C – Existing party structures

Scope of rights of building owner

Under section 2 of the 1996 Act, a building owner has a statutory right to interfere with the proprietary rights of the adjoining owner. The building owner might wish to resort to these rights by rectifying defects, want of repair, renewal before repair becomes necessary, or rebuilding with more durable materials. Section 2,

---

[13] *Selby* v *Whitbread & Co* [1917] 1 KB 736 at 752 (McCardie J) (in relation to London Building Act 1894).
[14] *Frederick Betts Ltd* v *Pickfords Ltd* [1906] 2 Ch 87.
[15] An adjoining owner may, by long inactivity in the face of acts of possession by his neighbour, lose his right to claim that a wall is a party wall under the Limitation Act 1980: *Prudential Assurance Co* v *Waterloo Real Estate Inc* The Times, May 13 1998.
[16] *Knight* v *Purcell* (1879) 11 ChD 412 at 414 (Fry J).
[17] *Drury* v *Army and Navy Auxiliary Co-Operative Society Ltd* [1896] 2 QB 721.
[18] *London, Gloucestershire and North Hants Dairy Co* v *Morley and Lanceley* [1911] 2 KB 257.

altering the previous rules,[19] confers on a building owner the right to demolish a party wall or a party fence wall and rebuild it to a reduced height, within definined limits (s 2(2)(m)). These are, notably, to a height of not less than two metres where the wall is not used by an adjoining owner to any greater extent than a boundary wall.

Party structure notice

A building owner cannot exercise any of the rights prescribed by section 2 of the 1996 Act unless he has first complied with a notice procedure. The procedure appears to be mandatory. The High Court compelled a defendant who had demolished a party wall without having previously notified the adjoining owner, to discontinue all further work and to remove the building work from the latter owner's land.[20] In this case, the proceedings were for interlocutory relief. A plaintiff who was guilty of unreasonable delay in pursuing his rights might fail to obtain injunctive relief but these claimants acted promptly. Although the 1996 Act does not in terms specify that statutory notices are to be in writing, the scheme of the Act by necessary implication requires that written and not oral notices should be given.[21]

By section 3(1), before exercising any right conferred on him by section 2 of the 1996 Act, a building owner must serve on the adjoining owner a "party structure notice", stating notably the nature and particulars of the proposed work, and the date at which it will be begun. If the building owner proposes to construct special foundations, his particulars must include plans, sections and details of the construction of these with reasonable particulars of

---

[19] See *Gyle-Thompson v Wall Street (Properties) Ltd* [1974] 1 All ER 295 (concerned with s 46 of the London Building Acts (Amendment) Act 1939).

[20] *London & Manchester Assurance Co Ltd v O & H Construction Ltd* [1989] 2 EGLR 185

[21] As noted by the Current Law Statutes Annotated commentary on the 1996 Act, the word "other document" appears in s 15 and in addition s 15(1) envisages delivery of a notice to an adjoining owner or service by post. Moreover, s 10 specifies that appointments of surveyors must be in writing and s 15 is presumably intended to be consistent with this provision. For a specimen party structure notice, see Bickford-Smith and Sydenham p 162–163.

the loads to be carried. "Special foundations" mean, by section 20, foundations in which an assemblage of steel beams or rods is employed for the purpose of distributing any load.

By section 3(2)(a) of the 1996 Act, a party structure notice must be served at least two months before the date on which the proposed work will begin. Unless within twelve months after the day of service of a party structure notice,[22] the work to which it relates is not begun and prosecuted with due diligence, the notice ceases to have effect (s 3(2)(b)).[23] Once a party structure notice has been served, any adjoining owner on whom it has been served has 14 days in which to serve a notice indicating his consent to the party structure notice. If he does not avail himself of this right, he is deemed to have dissented from the notice. A dispute is deemed to have arisen between the parties (s 5). The statutory disputes resolution procedure under section 10 then comes into being. One seeming aim of the Act is the rapid resolution of disputes.

Counter-notice

An adjoining owner has the right to serve a counter-notice on the building owner: this notice must be served within one month of the day of service of the party structure notice (s 4(2)(b)).[24] This notice must, in particular, specify the works required by the notice and be accompanied by plans, sections and particulars of the works (s 4(2)(a)). In addition, a counter-notice may require the building owner to build, in relation to a party fence wall or party structure, such chimney copings, breasts, jambs, or flues or such piers, recesses or other like works, as may reasonably be required for the convenience of the adjoining owner (s 4(1)(a)). If the adjoining owner consents to special foundations, his counter-notice may

---

[22] To avoid nullification of statutory arbitrations, if a surveyors' award is made, a dispute having arisen, only outside the 12 months, this provision does not apply: *Leadbetter* v *Marylebone Corporation* [1905] 1 KB 77, CA (in relation to the former period of six months).
[23] However, a building owner may exercise the rights conferred on him by s 2 with the consent in writing of the adjoining owners and occupiers (s 3(3)).
[24] It has been recognised judicially that the time for service of a counter-notice is in practice frequently extended by agreement: *Chartered Society of Physiotherapy* v *Simmons Church Smiles* [1995] 1 EGLR 155 at 157.

require these to be placed at a specified depth greater than that proposed by the building owner (s 4(1)(b)(i)), or that the foundations be constructed to a sufficient depth to bear the load to be carried by columns of any intended building of the adjoining owner (s 4(1)(b)(ii)), or both. If the building owner wishes to reduce the height of a party wall or party fence wall (under s 2(2)(m)) the adjoining owner has the right to serve a counter-notice under section 11(7) requiring him to maintain the existing height of the wall.[25]

Section 4(3) obliges the building owner to comply with a counter-notice, unless the execution of the works would be injurious to him or would cause unnecessary inconvenience or unnecessary delay in the execution of the works he proposes in his party structure notice.[26] If there is more than one adjoining owner (such as a lessee other than from year to year) then a notice of intent must be served on all or any of them.[27] The building owner has a 14-day period in which to serve a notice consenting to the counter-notice and if he does not, then he is deemed to dissent to it. A dispute is deemed to exist between the parties (s 5) with the result that the disputes resolution procedures of the Act come into play.

Rights of building owner under section 2 – Introduction

By section 2(1) of the 1996 Act, where at the line of junction between adjoining "lands" of different owners the "lands" are built on or there is a boundary wall which is a party fence wall or the external wall of a building, then section 2(2) of the 1996 Act confers on the building owner a list of rights.[28] These are subject to various qualifications in certain cases in the interests of the adjoining owner. The exercise of these rights is set out in a comprehensive way and to that extent section 2 supercedes the common law.

---

[25] In such a case the adjoining owner must bear a proportion of the cost of the additional work so caused (s 11(7)).
[26] A safeguard for an adjoining owner is the right (s 12) to require the building owner to give security for the costs of the work proposed, as agreed or resolved under s 10.
[27] *Crosby* v *Alhambra Co Ltd* [1907] 1 Ch 295.
[28] There is also an obligation under s 6 of the 1996 Act on a building owner who is building within a prescribed distance of an adjoining owner's land, where excavations are involved, to serve a notice of intent on that owner.

Compensation

The building owner concerned must compensate both any adjoining owner and occupier for any loss or damage which may result to any of them by reason of any work executed in pursuance of the Act (s 7(2)). The works must comply both with any relevant statutory requirements and be executed in accordance with plans, sections and particulars agreed or determined under the statutory disputes procedure (s 7(5)). There is a wide right of entry which, save in the case of emergency, requires 14 days' notice to the adjoining owner and occupier of the premises. It is conferred on the building owner, his servants, agents or workmen during usual working hours for the purpose of executing work in pursuance of the Act (s 8(1)).[29]

If a building owner carries out works to a party structure in such a way as to constitute a nuisance to the adjoining owner, but fails to comply, before starting the works, with the procedures of the legislation, he cannot rely on the legislation to safeguard him from his usual common law liabilities, or even to reduce them. Thus, where a person who was in fact a "building owner" within the 1939 legislation carried out works which interfered with his neighbour's party wall, and only on completion of the works did he serve the requisite statutory notices, he was liable in damages to the adjoining owner.[30] The award included special damages from the date their premises would have been sold, but for the damage caused by the works, until repairs had been completed, in respect of mortgage interest, so illustrating the perils of neglecting prompt compliance with the statutory rules and procedures.

Rights of building owner

The 1996 Act confers various rights on a building owner, the principal ones being as follows:

---

[29] To avoid doubt, the power extends to the removal of furniture or fittings for the purpose (s 8(1)). There is a power in the building owner and his servants etc to break into closed premises if accompanied by a police officer (s 8(2)). It is an offence for an occupier of land or premises to refuse to permit a building owner or other person such as his surveyor to do anything which he is entitled to do with regard to the land or premises, provided the occupier knows or has reasonable cause to believe that the person is so entitled (s 16).

[30] *Louis* v *Sadiq* [1997] 1 EGLR 136.

1. To underpin, thicken or raise a party structure, a party fence wall or an external wall which belongs to the building owner and is built against a party structure or party fence wall (s 2(2)(a)). If the work mentioned is necessary on account of defect or want of repair to the structure or wall concerned, the expenses must be apportioned between the building owner and the adjoining owner in accordance with section 11(4) discussed above. As a deterrent to unnecessary works, if the work is not required to cure a defect or want of repair, then by section 2(3) of the 1996 Act, the right to carry out the work is, in particular, subject to the building owner making good all damage occasioned by the work to the adjoining premises or to their internal furnishings and decorations.[31]
2. To make good, repair, or demolish and rebuild, a party structure or party fence wall where such work is necessary on account of defect or want of repair in the structure or wall (s 2(2)(b)).[32] In relation to an earlier, differently worded, provision,[33] the House of Lords held that the right of a building owner to make good a defective party structure or one which was out of repair was confined to making good the party structure so that it became effective for the purpose for which it was actually used or intended to be used. Their Lordships also ruled that in deciding on the proper way to make good a defect, the court was entitled to have regard to the convenience of the adjoining owner.[34] Thus dampness in a party wall was not a defect within the Act unless it made the wall less effective for the purposes for which it was used or intended to be used. In that case, the repairs to cure damp could be carried out without any intrusion by the building owner into the adjoining owner's premises and the claim by the former to enter those premises was accordingly dismissed. There is no reason why these rulings would not

---

[31] In addition, s 2(3)(b) requires in the case of work to a party structure or external wall, the carrying up of any flues and chimney stacks to an agreed height and with agreed materials (the disputes procedure applies if there is no agreement).

[32] A right of contribution from the adjoining owner is conferred by s 11(5) of the 1996 Act.

[33] London Building Act 1894 s 88(1), which referred to "a right to make good underpin or repair any party structure which is defective or out of repair".

[34] *Barry v Minturn* [1913] AC 584 at 590 (Lord Parker of Waddington).

apply today despite the narrower language of the provision then being considered.

3. To demolish a partition which separates buildings belonging to different owners but does not conform with statutory requirements and to build a party wall which conforms (s 2(2)(c)).[35] However, for the avoidance of doubt, a building or structure which was erected before July 18 1996 is deemed to conform to statutory requirements if it conformed with statutes regulating buildings or structures on the date it was erected (s 2(8)).
4. To demolish a party structure which is of insufficient strength or height for the purposes of any intended building of the building owner and in particular to rebuild it of sufficient strength or height for these purposes (s 2(2)(e)).[36]
5. To cut into a party structure for any purpose (which may be or include inserting a damp-proof course) (s 2(2)(f)).[37] Similarly, to cut into the wall of an adjoining owner's building in order to insert a flashing or other weather-proofing of a wall erected against that wall (s 2(2)(j)).[38]

---

[35] There is also a related right of demolition and rebuilding in the case of the whole or part of buildings, arches or structures over public ways or belonging to other persons, where the originals do not comply with statutory requirements (s 2(2)(d)).

[36] This right, which is qualified by s 2(4) so as, in particular, to require the making good of all damage occasioned by the work to the adjoining premises and to their internal furnishings and decorations, includes rebuilding to a lesser height or thickness where the rebuilt structure is of sufficient strength and height for the purposes of the adjoining owner (presumably as at the date of the work and not his future purposes).

[37] This right is subject to an express obligation (s 2(5)) to make good all damage occasioned by the work similar to that of s 2(4), a qualification also applying to rights to cut away from a party wall, party fence wall etc any footing or any projecting chimney breast, jamb or flue or any other projection on or under the land of the building owner in order to erect, raise or underpin any such wall or for any other purpose (s 2(2)(g)) and to cut away or demolish parts of any wall or building of an adjoining owner which overhang the land of the building owner or a party wall to the extent necessary to enable a vertical wall to be erected or raised against the wall or building of the adjoining owner (s 2(2)(h)).

[38] This right is subject to a duty in the building owner to make good all damage occasioned by the work to the wall of the adjoining owner's building (s 2(6)).

6. A right is conferred on the building owner to execute other necessary works incidental to the connection of a party structure with the premises adjoining it (s 2(2)(k)).
7. A right is also conferred on him to raise a party fence wall and to demolish a party fence wall and to rebuild it as a party fence wall or as a party wall (s 2(2)(l)).
8. Section 2(2)(m) of the 1996 Act, which is new,[39] allows a building owner to reduce or to demolish and rebuild a party wall or party fence wall to a specified minimum height.[40] While a building owner has now the right to expose a party wall or party structure hitherto enclosed, he must, clearing up difficulties at common law,[41] provide adequate weathering.

D – New party structures

Where a building owner wishes to build a party wall or a party fence wall on the line of junction where it is not currently built on, or the line of junction is built on only to the extent of a boundary wall which is not a party fence wall within the Act or the external wall of a building, he falls within section 1. At least one month's notice before the intended date of the start of the work is required from the building owner to the adjoining owner, stating his desire to build and describing the intended wall.[42] If the adjoining owner serves a notice of consent on the building owner,[43] the wall will be

---

[39] It is subject to s 2(7): the building owner must reconstruct or replace any parapet. If a parapet is now needed, he must construct one.
[40] Or to a height currently enclosed upon by the building of a building owner.
[41] Caused by *Phipps v Pears* [1965] 1 QB 76, CA, and despite the rulings in *Upjohn v Seymour Estates Ltd* [1938] 1 All ER 614.
[42] A requirement of a notice of intent to the adjoining owner describing the wall is imposed even where the building owner wishes to build on the line of junction a wall placed wholly on his own land (s 1(5)). Where he so builds, and also where he builds a wall on his own land having failed to obtain the adjoining owner's consent to building partly on the latter's land, the building owner has the right to place below the land of the adjoining owner any necessary projecting footings and foundations within a 12-month period from service of his original notice of intent (s 1(6)).
[43] ie a notice served (in writing) within 14 days beginning with the date of service of the building owner's notice (s 1(4)).

built half on the land of each of the two owners – or in such other position as may be agreed. The expenses of building the wall are to be shared out between the owners, in proportion to the use made by each of them of the wall and to the cost of labour and materials prevailing at the time when that use is made by each owner respectively (section 1(3)). If the adjoining owner does not consent, the wall must be built by the building owner at his expense, wholly on his own land, and as an external wall or a fence wall (s 1(4)). Section 1 does not authorise the obstruction of ancient lights.[44] Any disputes between the owners are to be dealt with under the specific section 10 procedures.

E – Resolution of disputes

Where a dispute arises between owners or is deemed to exist, the matter must be settled within the procedures of section 10 by surveyors. The parties may concur in the appointment of one surveyor – the agreed surveyor. Or each party must appoint one surveyor and they must forthwith select a third surveyor (s 10(1)). All appointments and selections must be in writing and cannot, once made, be rescinded by either party (s 10(2)). It appears that the policy of this provision is to prevent a party from avoiding an award being made by an early discharge of his surveyor.[45]

The Act aims to promote the speedy resolution of disputes. Thus, if an agreed surveyor refuses or neglects to act for ten days after a written request from either party to act, or before the difference is settled he dies, or becomes or deems himself incapable of acting, all proceedings to settle the dispute must begin *de novo* (s 10(3)). If either party refuses by notice to appoint a surveyor, or for ten days after a written request from the other party neglects to do so then the other party may make the appointment on his behalf (s 10(4))). Under section 10(6), if a surveyor, however appointed, refuses to act effectively, the surveyor of the other party may proceed to act *ex parte*.[46] Everything so done by him is as effectual

---

[44] *Crofts v Haldane* (1867) LR 2 QB 194.
[45] Anstey, *Trouble with the Neighbours*, p 14.
[46] There are default provisions where a surveyor neglects to act effectively (s 10(7)); also as to where a surveyor refuses or neglects to appoint a third surveyor (s 10(8)) and where a third surveyor refuses or neglects to act (s 10(9)) – no use is there made of the term "effectively".

as if he had been an agreed surveyor. Thus, he is then able to draft and make an award by himself.

Usually, two agreed surveyors, or the three surveyors or any two of them will make their award under section 10(10). Either party or either of their appointed surveyors may call upon the third surveyor, where relevant, to determine the disputed matters and he is then bound to make the necessary award (s 10(11)). An award may determine:

(a) the right to execute any work;
(b) the time and manner of executing any work;
(c) any other matter arising out of or incidental to the dispute including the costs of making the award.

However, the time referred to may be no earlier than the date of expiry of any statutory notice related to the dispute – unless the parties to the dispute otherwise agree (s 10(12)). Only once an award is made may work begin.

The power of a third surveyor to make an award is limited, despite the wide expression "any other matter" since he may only make an award determining whether one of two adjoining owners in dispute is permitted to carry out works under the Act and if so, as to the terms and conditions of the work. Thus, a third surveyor, under the 1939 Act rules, was held to have no jurisdiction to make an award which purported to determine the prime cause of damage which had caused a party wall to collapse.[47] There is no reason why this construction should not apply to the 1996 Act. The 1996 Act provides that an award is conclusive (s 10(16)). However, either party has the right within 14 days of service on him of an award to appeal to the county court (s 10(17)).[48] The county court has wide powers to rescind or modify the award.

It was held in the Official Referee's court in relation to the predecessor to section 10 (which was cast in similar, though not in all respects identical, terms) that a statutory award was *sui generis* and was an expert determination rather than an arbitration. The

---

[47] *Woodhouse* v *Consolidated Property Corporation* [1993] 1 EGLR 174 at 177C, CA.
[48] ie from the date when the owner takes up an award after notification of it by letter: *Riley Gowler Ltd* v *National Heart Hospital Ltd* [1969] 3 All ER 1401, CA.

court may completely reopen an award on appeal, subsituting its findings of fact for those of any surveyor. The court does not simply ask itself the narrow question, appropriate in the context of the Arbitration Act 1979, which is excluded by necessary implication, of whether the award was such that it could have been made by a competent surveyor: it can thus receive admissible evidence not available to the surveyor.[49]

Challenges to awards

Surveyors, in making an award, have no power to interfere, save to the extent allowed by the 1996 Act, with existing property rights in the party wall of an adjoining owner. If the award does this, it may be challenged outside the statutory time-limits for an appeal.[50] A similar challenge to an award, again outside the statutory time-limits, may be made to an award which is said to be invalid for want or excess of jurisdiction.[51]

Invalidity of award for non-compliance with procedures

The approach of the parties and their surveyors to the procedural requirements of the 1996 Act must of necessity be meticulous, as the High Court ruled that a surveyors' award may be invalid on procedural grounds alone.[52] Such potentially fatal procedural irregularities could include non-service of a requisite notice, or service on the wrong party, as where a party structure notice is served not on the owner concerned, but on his surveyor (where the latter is not authorised expressly or impliedly to accept service). Similar procedural irregularity would arise where agreed surveyors or the third surveyor are not appointed in writing, as required by section 10.

---

[49] *Chartered Society of Physiotherapy* v *Simmons Church Smiles* [1995] 1 EGLR 155.
[50] *Burlington Property Co Ltd* v *Odeon Theatres Ltd* [1938] 3 All ER 469, CA.
[51] *Gyle-Thompson* v *Wall Street (Properties) Ltd* [1974] 1 All ER 295 (though reversed on the facts by s 2 of the 1996 Act).
[52] *Gyle-Thompson* v *Wall Street (Properties) Ltd, supra.*

III – Dangerous buildings or structures

Local authorities have powers to control dangerous buildings or structures under section 77 of the Building Act 1984.[53] Under section 77(1), if it appears to a local authority that a building or structure (or part thereof) is in a dangerous condition, the authority may apply to a magistrates' court. The court then has power to order the owner to carry out any remedial work necessary to obviate the danger or, at his election, to demolish the building or structure (or any dangerous part of it). In the event of the owner failing to comply with a court order then (s 77(2)) the local authority have the power to execute the order in such manner as they think fit and to recover the expenses reasonably incurred by them from the owner. Moreover, non-compliance with a court order is an offence. If the building is a listed building, then a local authority considering demolishing the building must first consider whether any alternative course of action is open to them to prevent demolition taking place.[54]

A – Emergency action

If it appears to the local authority that immediate action should be taken to remove a danger caused by a dangerous building or structure or part of either, they have emergency powers under section 78. They must if reasonably practicable to do so, give notice of intent to the owner and occupier (s 78(2)). In principle, the authority have a right to recover expenses they incur (s 78(3)). By section 78(5), the court, in any proceedings against the owner to recover expenses, must inquire whether the local authority might reasonably have proceded under section 77(1). If it decides that they should have done so, the authority will lose their right to recover any expenses.

---

[53] For separate but in most respects similar rules in Inner London see Part VII of the London Building Acts (Amendment) Act 1939 (which was not repealed by Party Wall etc Act 1996 (Repeal of Enactments) Order 1997, SI 1997 No 671). Section 79 of the 1984 Act empowers local authorities to require the owner of a dilapidated or dangerous building to repair, restore or demolish it, at his election, where its condition is seriously detrimental to the amenities of the neighbourhood.

[54] See *R v Stroud District Council* [1982] JPL 246, where mandamus was obtained against an authority which declined to consider listing.

B – Further considerations

The fact that the local authority temporarily shore up a structure will not prevent it being in a dangerous state, and the owner will still be liable to be proceeded against under section 77 (or, presumably, under s 78). Apprehension of danger to any person, not just to highway users, suffices for the statutory power to be available.[55] Moreover, a court order requiring the owner to carry out remedial work to a dangerous building or structure does not necessarily have to specify the works, and may simply refer to "such works of repair or restoration" as may be required to comply with the order.[56] As to what work is necessary to render the building or structure safe, it has been held that it is work of a semi-permanent nature which would make the building or structure reasonably safe in respect of any person who might happen to go into it[57] If a local authority cause a dangerous building or structure to be demolished and, on one side of the building there is a wall dividing it from other premises, if the wall is a party wall, the owner of the demolished building may well have to pay for the costs of supporting the party wall.[58]

---

[55] *London County Council* v *Jones* [1912] 2 KB 504.
[56] *R* v *Recorder of Bolton* [1940] 1 KB 290, CA.
[57] *Holme* v *Crosby Corporation* (1941) (unrep), cited in Current Law Statutes notes to Building Act 1984, s 77.
[58] See *Marchant* v *Capital & Counties plc* [1983] 2 EGLR 156, CA.

Chapter 12

Agricultural dilapidations

I – Introduction

There are two sets of statutory rules now applying to agricultural land and premises. In the case of any tenancy granted before September 1 1995, the governing statute is the Agricultural Holdings Act 1986. To that regime apply, unless excluded in terms, the model clauses, which are the product of statutory instrument.[1] These are constructed, it is said, on the basis of liability under an annual periodic tenancy with (potential) lifetime security.[2] In the case of a tenancy of agricultural land and premises granted on or after September 1 1995, the governing statute is the Agricultural Tenancies Act 1995, which creates "farm business tenancies". Under the 1986 Act, the parties will be bound by the model clauses regime unless they expressly agree to be subject to the rules of the common law, itself subject to the damages limitation provisions of section 18(1) of the Landlord and Tenant Act 1927. The onus on the landlord is to prove loss from the tenant's alleged breaches of repairing obligation. If he fails to produce evidence of breach, especially in the case of a breach of covenant at the beginning or in the mid-term of a long lease, the landlord runs the risk that the court will assume that he has suffered no significant diminution to the value of his reversion.[3]

By contrast to the 1986 Act, the 1995 Act was passed to promote "deregulation", so that there is no security of tenure conferred on a farm business tenant. In principle, farm business tenancies are subject to common law principles, subject to the 1927 Act ceiling on damages. However, it is open to the parties to a farm business

---

[1] Agriculture (Maintenance Repair and Insurance of Fixed Equipment) Regs 1973, SI 1973 No 1473, as amended by SI 1988 No 281.
[2] Scammell and Densham, p 973.
[3] *Crewe Services & Investment Corporation* v *Silk* [1998] 2 EGLR 1, CA. Equally, the court was not bound to award nominal damages to a landlord who proved no loss.

tenancy to incorporate the model clauses into their agreement: if so, it should be specified whether the regulations as originally enacted or as amended are intended to bind the parties.[4]

As seen, in case of tenancies to which the Agricultural Holdings Act 1986 applies, the terms of any written tenancy may make express provisions allocating liability for particular repairs and maintenance of the fixed equipment of the holding, such as the farmhouse, cottages and any farm outbuildings.[5] In this case the terms of the tenancy are a matter of contractual interpretation. Otherwise, as where a written tenancy agreement makes no express provision as to repairs of fixed equipment, or where the tenancy is oral, or where there are gaps in any express provisions of the tenancy, the model clauses are automatically incorporated into every contract of tenancy of an agricultural holding (s 7(3)).

Both the landlord and the tenant have a right to refer a written agreement under the 1986 Act to an arbitrator, if it makes substantial modifications to the 1973 model clauses. Unless a right to refer the terms of a written tenancy is exercised, its express terms will continue to prevail.[6] In order to be valid, a reference to arbitration must be in writing and must follow an unsuccessful written request to the other party to vary the agreement so as to bring it into conformity with the model clauses (1986 Act, s 8(2)). The powers and discretions of the arbitrator are wide. He is bound to consider whether the express terms of the written agreement are justifiable having regard to the circumstances of the holding and of the parties (s 8(3)). If he determines that the disputed terms are not justifiable, he may vary them in such manner as he considers reasonable and just. The arbitrator is empowered to vary the rent of the holding if it is equitable to do so. Once a reference has been made to an arbitrator, there is a three-year time-bar on the making of a further reference of the terms of the tenancy concerned, as from the coming into effect of his award in relation to the initial reference (s 8(6)).

---

[4] Scammell and Densham, pp 972–973, who draw attention to the fact that even if the model clauses as amended are referred to in the agreement, if the regulations are further amended, the latest set of amendments will not, without specific provision, be included. Also Muir Watt and Moss, para 10–11.
[5] "Fixed equipment" is defined for the purposes of the 1986 Act by s 96(1) and is extended by SI 1973 No 1473, reg 1.
[6] *Burden v Hannaford* [1956] 1 QB 142, CA.

II – Liability for repairs to buildings and fixed equipment under the rules of the 1986 Act

A – Principles of allocation of liability

As already seen, whether or not the parties have agreed on written terms governing to a complete or incomplete extent, the liability for repairs will be subject to modification by an arbitrator in the light of the model clauses. These apply in any case to tenancies which are silent as to the allocation of liability for repairs. It has been recognised judicially that the model clauses were enacted for the protection of agricultural tenants, by placing clear liabilities to repair and replace on the landlord. By means of the procedure, discussed below, of a tenant's default notice, the tenant has a speedier remedy than he would obtain by action in the courts, seeing that any disputes as to such notices must be referred to arbitration.[7]

Hence, the model clauses allocate the liability for major structural repairing and maintenance work to farm buildings and fixed equipment to the landlord, who is entitled to recover half the cost of smaller maintenance items. The tenant is liable for such matters as interior repairs and redecoration to farm buildings. The regulations do not affect the interpretation of expressions such as "repair" or "renew", which are governed by common law principles discussed elsewhere in this book. The regulations provide for arbitration as to the extent of liability, seeing that the landlord does not have to carry out improvements as opposed to repairs, and as to whether a breach of obligation has taken place. An agricultural tenant is entitled to claim specific performance against his landlord, at the discretion of the court, once, but not until, liability to carry out particular work has first been established or presumed against the landlord, following an arbitration award.[8]

B – Allocation of liability by model clauses

Landlord's liabilities

The main responsibilities of the landlord under the model clauses are as follows. He is liable to carry out repairs and replacements to

---

[7] *Hammond* v *Allen* [1994] 1 All ER 307.
[8] *Tustian* v *Johnston* [1993] 2 EGLR 8. An arbitrator has no such power: *ibid*.

the structure and exterior of the farmhouse, cottages and farm buildings (including, for example, the roofs, chimney stacks, main and exterior walls).[9] He must make good any interior repair or decorations made necessary as a result of structural defects to such roof, walls and so on (para 1). The landlord must insure the farmhouse and other farm buildings against loss or damage by fire to their full value and is obliged by the regulations to reinstate these if the event materialises (para 2). As might be expected, the landlord is obliged to carry out decorative work to the outside wood- and iron-work of the farmhouse, cottages and farm buildings, and to the inside wood- and iron-work of all external outward-opening doors and windows of farm buildings, as often as is needed to prevent deterioration, and in any case at least once every five years, to the standard specified in the regulations (para 3(1)). He may recover half the reasonable cost from the tenant for doors, windows, eaves-guttering and downpipes (para 3(1)), with a proportionate reduction in the sum recoverable if the work is carried out before the fifth year of the tenancy.

If the landlord fails to carry out any repairs for which he is responsible (there is a special rule for repairs to an underground water pipe in para 12(2)) within three months of receiving a written notice from the tenant, specifying the repairs and calling on him to execute them, the tenant may execute the work and recover the reasonable cost from the landlord forthwith (para 12(1)). The landlord may contest the tenant's notice by a one-month counter-notice, time being of the essence in relation to its service, as the limit is for the tenant's benefit. Otherwise, liability will be presumed against the landlord.[10]

---

[9] According to Muir Watt and Moss, para 10–12, this "curiously drafted" regulation is intended to include floors, floor joists, ceiling joists and timbers, exterior and interior staircases and fixed doors, ladders, windows and skylights. It is to be noted that, for the protection of the landlord against surprise claims, the regulations require the tenant to report any damage to the landlord in writing as and when it occurs. The tenant is bound to protect all items the landlord is liable to repair from wilful, reckless or negligent damage (para 5(4)) and if he fails to do so, the landlord is not liable for repairs to the items concerned (para 4(1)(b)).

[10] *Hammond* v *Allen* [1993] 1 EGLR 1. The matter then goes to arbitration (para 12(5)).

A right of recovery, following a three-month notice, and execution of the works by the tenant, is given to the tenant in the case of failure by the landlord to execute replacements for which he is liable (para 12(3)).[11] In this case, however, the tenant is limited to recover the annual rent of the holding or, in respect of claims in any year of the tenancy terminating as from March 24 1988, of £2,000, whichever is the smaller sum.[12] The tenant's right to recover against the landlord under the model clauses has been held to be exclusive of any right to pursue a damages claim, as the model clauses provide a self-contained scheme. He is, however, entitled to recover up to the limit of the sums specified in the regulations for each year of the tenancy until the whole claim is exhausted, with appropriate adjustments in respect of replacements carried out by the landlord in that year.[13]

Tenant's liabilities

Certain minor work of repair and maintenance to the farmhouse, cottages and farm buildings and to fittings and fixtures above ground and other specified items therein[14] fall on the tenant, which premises or items he is bound to repair and to keep and leave clean and in good tenantable repair, order and condition[15] save to the extent that the landlord is obliged to repair and maintain them (para 5(1)). The tenant has, in particular, to renew and pay for the cost of the replacement of broken or cracked tiles or slates and to replace all slipped tiles or slates, as and when damage occurs, but only up to an annual limit for any one year of the tenancy of £25 until March 24 1988 and £100 as from that date (para 8). Subject to the landlord's responsibilities, the tenant must carry out repairs and replacements and, in the process, decorations to fixed equipment and must paint or (as appropriate) treat such equipment

---

[11] The landlord may contest the tenant's notice by a counter-notice whose effect is suspensory of the tenant's notice pending statutory arbitration.

[12] In respect of claims in any year of the tenancy terminating before March 24 1988, as from which date the 1988 amendment regulations commenced, the fixed financial ceiling is £500.

[13] *Grayless* v *Watkinson* [1990] 1 EGLR 6, CA.

[14] Including therefore electical wiring: *Roper* v *Prudential Assurance Co Ltd* [1992] 1 EGLR 5; also expressly including all fences, hedges, fields, walls, gates, ponds, ditches and yards.

[15] As to the standard so imposed see *Evans* v *Jones* [1955] 2 QB 58, CA.

with effective preservative material (para 6(1)). He must also carry out decorative work such as painting to the interior of the farmhouse, cottages and farm buildings as often as may be necessary and at least once every seven years and in the last year of the tenancy(para 7). Special provision is made to enable the landlord to recover from the tenant sums in respect of accrued liability to carry out interior repairs where the tenancy terminates in a year within the seven-year interval (para 11(1)).

The landlord may serve a three-month notice on the tenant, it being assumed that time is of the essence, requiring him to execute the repairs for which he is liable, with a default power of entry and execution (para 4(2)).[16] Alternatively, the landlord may serve a notice to remedy on the tenant.[17] Should this notice not be complied with, the landlord may in due course serve an uncontestable notice to quit on the tenant. But the tenant may contest the notice to remedy by referring the notice to an arbitrator, who has powers to modify it.[18] However, an advantage of a notice to remedy is that, except and to the extent that it is modified, the tenant must fully comply with it, within the reasonable time allowed for a remedy of each breach alleged.[19] It is no defence to liability for the tenant to claim that he cannot remedy the breach because the landlord has failed, in breach of covenant, to supply him with necessary materials.[20]

Because the policy of the regulations is to limit the tenant's obligations to routine repairs, it is the landlord who must replace items which have become worn out or otherwise incapable of further repair (para 1(3)), unless the condition of the item is the result of the tenant's own fault.

---

[16] The tenant may contest this notice and the dispute then goes to arbitration (para 4(3)). Time is seemingly of the essence.
[17] Agricultural Holdings Act 1986, Sched 3 Part I Case D, in the form prescribed by SI 1987 No 710, para 3. The landlord may opt to ask the Agricultural Land Tribunal for a certificate of bad husbandry (1986 Act, Sched 3, Part III, para 9), after the issue of which a Case C notice may be served, which cannot be contested.
[18] The tenant may invoke s 28 of the 1986 Act so as to require the Agricultural Land Tribunal to consent to the Case D notice.
[19] See *Wykes v Davis* [1975] QB 843, CA.
[20] *Shepherd v Lomas* [1963] 1 WLR 962, CA.

III – Damages claims by landlord under the 1986 Act

Where a landlord has claims for dilapidations against the tenant of an agricultural holding, he may invoke various remedies, such as forfeiture, or apply for a certificate of bad husbandry or serve a notice to do work (as to the latter see above). But he may elect to claim damages under sections 71 and 72 of the 1986 Act. If the current tenant obtains a new tenancy of the same, but not an altered, agricultural holding, the landlord preserves his right to claim compensation against the tenant under sections 71 and 72 (s 73). Should the landlord resume possession of a severed part of the holding, his rights to claim compensation against the tenant with respect to that part of the holding are also preserved (s 74).

A – *Claim under section 71 for deterioration to particular parts of holding*

Claims under statute

Section 71(1) of the 1986 Act entitles the landlord to recover from the tenant, "compensation in respect of the dilapidation or deterioration of, or damage to, any part of the holding or anything in or on the holding", caused by the tenant's non-fulfilment of his responsibilities to farm in accordance with the rules of good husbandry. A claim by the landlord under section 71(1) is a statutory claim and, since it may only be made on the tenant's quitting the holding, it must go to arbitration under the 1986 Act (s 84) if the parties cannot agree on the settlement of the claim. Such claims cannot be pursued in the courts.

The word "dilapidation" is widely understood and includes claims for disrepair of particular buildings, neglect of fences or hedges, the repair of bent or broken gates, ditches or drains and culverts, the repair or maintenance of bridges and the fouling of land.[21] But claims for general dilapidations are to be made under section 72.

In assessing whether the landlord proves his claim, the arbitrator must take into account the express terms of the tenancy, if any, the terms of any other agreement affecting the holding and of the 1973 model clauses (so far as incorporated into the tenancy) and also the

---

[21] *Evans v Jones* [1955] 2 QB 58, CA.

duties to farm in accordance with the rules of good husbandry.[22] The standard of repair imposed on the tenant in respect of buildings and other fixed equipment by the model clauses, for example, is that to "keep in repair and leave clean and in good tenantable repair, order and condition" the specified items. It appears that the tenant's obligation is not merely to keep and leave those things as clean and in as good repair as they happened to be at the beginning of the tenancy, but equally, in deciding whether the obligation has been complied with with respect to any item, regard should be had to its age and character and its condition at the commencement of the tenancy. All the circumstances are relevant factors, including the condition of the items, their age and the length of the tenancy.[23]

Compensation under section 71(1) may only be claimed on the tenant's quitting the holding on the termination of the tenancy (s 71(4)(a)), whether voluntarily or following a notice to quit. The amount of the compensation is the cost, at the date of the tenant's quitting the holding, of making good the dilapidation, deterioration or damage (s 71(2)). There is a ceiling on the maximum amount of the damages, namely, the amount by which the value of the landlord's reversion is diminished by the breach in question (s 71(5)).

Claim under contract

Section 71(3) provides that the landlord may, instead of claiming compensation under section 71(1), claim compensation in respect of matters specified in that subsection in accordance with a written contract of tenancy. Such a claim might be advantageous to the landlord where a contract of tenancy makes different and more onerous provisions to those of the model clauses, but the tenant could then presumably seek an arbitration so as to modify the terms of the tenancy agreement.

---

[22] *Evans v Jones,* supra. The latter rules are in Agriculture Act 1947 s 10(3) and 11(3), incorporated into the 1986 Act by s 96(3), as to which see further Muir Watt and Moss, para 13.82. They note that these rules, enacted immediately post-war, "frequently do not match the modern complaints of landlords and the modern activities of tenants". EU policy is seemingly to encourage intensive farming or none at all.
[23] *Evans v Jones, supra* at 66 (Evershed MR).

The claim is to be made for dilapidations to particular buildings or fixed equipment or to the land and it must be made only on the tenant's quitting the land. It must, as with a claim under section 71(1), be pursued before an arbitrator and not in the courts. As with claims under section 71(1), compensation payable under section 71(3) is not to exceed the amount by which the value of the landlord's reversion is diminished by the breach in question (s 71(5)). The upper limit imposed on damages recoverable by section 18(1) of the Landlord and Tenant Act 1927 also applies to claims under section 71(3). This seems curious. One explanation may be that what is now section 71(5) of the 1986 Act was added to section 71 only in 1984.[24] There is no reason, in the absence of clear language either way in the legislation, why the reference to "premises" imposed by section 18(1) of the 1927 Act should apply narrowly to farm buildings and fixed equipment alone, as opposed also to the farm land.[25]

Landlords' claims during tenancy

The landlord retains the right to claim damages, at common law, during a written contract of tenancy in respect of dilapidations or damage to particular buildings, fixed equipment or parts of the holding, rather than being forced in all cases to await the end of the tenancy and the departure of the tenant.[26] This right was preserved by necessary implication, since a statute is not to be construed as taking away a landlord's legal rights unless such an intention clearly appears. Under a long lease, a claim for dilapidations to particular parts could no doubt be made, where the tenant was alleged to be in breach of a general covenant to repair, apart from any failure by him to leave particular parts of the holding in repair on quitting the holding.

Further points

A restriction on claims to compensation, as opposed to damages, is

---

[24] By Agricultural Holdings Act 1984, Sched 3, para 13.
[25] For a narrow interpretation, see Muir Watt and Moss, para 13.94; but in its usual or popular sense, the word "premises" has been taken to comprise land and buildings: see eg *Whitley* v *Stumbles* [1930] AC 544, HL.
[26] *Kent* v *Conniff* [1953] 1 QB 361, CA.

that compensation cannot be claimed under section 71, on the termination of an agricultural tenancy, both under the statute and under the contract of tenancy (s71(4)(b)). The landlord must elect which claim to make, but he may frame alternative claims against the tenant and is only forced to his choice before the arbitration proceedings commence, seeing that he cannot claim twice over.[27] In any case, the onus of proving damage falls on the landlord and he may be assisted in this process where, at the commencement of the tenancy, a record of the condition of the holding and of the fixed equipment is made under section 22(1) of the 1986 Act. The landlord cannot compel the tenant to agree to this step.

Since claims under section 71(1) and (3) are made on the tenant quitting, these must be arbitrated upon, if the parties cannot agree on a figure of compensation. Hence, the landlord must give written notice to the tenant of his intention to claim compensation before the expiration of two months from the termination of the tenancy (s 83(2)) on pain of losing his right to claim arbitration. The nature of the claim must be specified in the landlord's notice, albeit in general terms, as by reference to the statutory provision relied on. With his notice claiming arbitration, the landlord may include a Schedule of Dilapidations and if the circumstances alter since the claim was made, the landlord has the right to ask the arbitrator to modify an existing head of claim.[28]

B – *Compensation for general deterioration*

In addition to any claim which he may have against the tenant for dilapidations or damage to particular buildings, fixed equipment or parts of the holding, the landlord may claim against the tenant, on his quitting the holding, for general deterioration of the holding under section 72 of the 1986 Act. If the landlord wishes to retain his right to make a claim, he must comply with section 72(4). The landlord must, therefore, not later than one month before the termination of the tenancy, give notice in writing to the tenant of his intention to claim this particular compensation. In addition, if the landlord wishes to preserve his right to claim arbitration, should the parties disagree as to a figure of compensation, he might well need to serve a separate written notice under section 83(2) of the

---

[27] *Boyd v Wilton* [1957] 2 QB 277, CA.
[28] See *ED & AD Cooke Bourne (Farms) Ltd v Mellows* [1982] 2 All ER 208, CA.

1986 Act on the tenant before the expiration of two months from the termination of the tenancy.[29]

The landlord must show that the "value of the holding generally has been reduced" owing to dilapidation, deterioration or damage or by non-fulfilment by the tenant of his responsbilities to farm in accordance with the rules of good husbandry (s 72(1)). In assessing whether the tenant is in breach of his duties, the arbitrator is to have regard to the written terms, if any, of the contract of tenancy, the terms, so far as incorporated into the tenancy, of the model clauses and the obligations to farm in accordance with the rules of good husbandry. Any compensation recovered by the landlord under section 71 for particular dilapidations must be brought into account in relation to a claim for general dilapidations under section 72 (s 72(2)), seeing that the landlord is not to be allowed to recover twice over.[30]

The amount of compensation recoverable for general deterioration or dilapidations is that equal to the decrease attributable to the matter in question in the value of the holding as a holding, having regard to the character and situation of the holding and the average requirements of tenants reasonably skilled in husbandry (s 72(3)). In effect, this limits the landlord to the drop, if any, in the value of his reversion occasioned by, for example, a breach by the tenant of his obligations to repair imposed by the contract of tenancy or the regulations. Subject to that ceiling, the arbitrator would no doubt compare the value of the buildings and other fixed equipment, at the date the tenant quit, with the value they would have had if the tenant had complied with his tenancy obligations and those imposed by the regulations.

Section 72 applies to any case where, as a result of specific failures by the tenant, which are the subject of claims under section 71, the landlord can also prove a general depreciation of his farm as a whole.[31] Such cases may occur in relation to the general running of the farm, rather than the repair and maintenance of individual buildings, as shown by the example suggested[32] of a tenant whose neglect of ditches causes general injury to field drainage.

---

[29] Despite *Hallinan v Jones* [1984] 2 EGLR 20: service of a second notice is a counsel of prudence (see Muir Watt and Moss, para 13.100).
[30] *Evans v Jones* [1955] 2 QB 58 at 64 (Evershed MR).
[31] *Evans v Jones, supra* at 64 (Evershed MR).
[32] In Muir Watt and Moss, para 13.99.

IV – Dilapidations and farm business tenancies

The Agricultural Tenancies Act 1995 creates a deregulated framework for farm business tenancies. Thus, in the case of repairing obligations, one might assume that the parties would be in the same position as any other parties in the commercial sector with regard to the negotiation and inclusion in the tenancy of express repairing obligations. It would not be likely that the courts would regard a farm business tenant as being any more deserving than a business tenant of the implication of any implied obligations against the landlord to repair or to keep in repair or in a fit state for use, even though it is understood that farm business tenancies are often of short duration, to that extent resembling assured shorthold tenancies, where tenants benefit from statutory intervention. The supposition would be that the parties are free to make whatever provision they like about repairs and to allocate liability as they think fit. If they fail to do so, on first principles, the court will not fill gaps. Equally, any express repairing obligations will be interpreted in accordance with normal common law rules, subject to the statutory ceiling on landlords' damages.

However, it is open to the parties to a farm business tenancy to incorporate expressly into the terms of the tenancy the model clauses. Such incorporation has the advantage that each party is governed by regulations which have been interpreted for some time and are in that sense well understood, but attention has been drawn to the disadvantages of this procedure.[33] One consists in the fact that if a tenancy expressly incorporates the 1973 regulations, it may fail to include these as amended in 1988. In any case, future amendments would not be covered unless such were clearly envisaged in the terms of the original tenancy. Moreover, the model clauses were not designed with a farm business tenancy in mind. Such tenancies are not automatically renewable, may be of short duration, and are sometimes aimed at encouraging a commercial undertaking (as shown by the possibility of diversification away from farming activities) rather than, under the 1986 Act, an agreement simply to farm tenanted land, possibly for life. Moreover, the disputes resolution procedures of the 1986 Act, a special form of arbitration thereunder, do not apply to the 1995 Act and any disputes about incorporated model clauses would presumably have to be referred to an arbitration under the general law.

---

[33] Scammell and Densham, pp 972–973.

Chapter 13

Third party rights and liabilities

I – Introduction

Third parties may claim in nuisance and negligence in respect of property damage or personal injuries caused by defects in dilapidated premises. These claims may be against the owner or occupier of the premises concerned. In this Chapter, we examine these principles and also a statutory right of access for repairs and related work onto the land of a neighbouring owner.

II – Liability of landlords in nuisance

Where premises have been let, it is not only the tenant, as occupier of demised premises, but also the landlord, who may be liable in nuisance on account of their dangerous state and condition.

A – Reservation of right to enter and repair

Where the premises have been let, and the lease expressly reserves a right in the landlord to enter and repair the demised premises, or where such a right is conferred on him by statute, he may be liable in nuisance to any third parties for damage or injuries caused by the defective condition of the premises. The landlord is ordinarily liable on the basis that he has control over the premises, either under a right to enter and repair or to inspect them, or because he is aware of the condition of the premises from his own inspections, or because he ought to have inspected them.

Where, therefore, a landlord let premises to yearly tenants, reserving himself a right to enter, inspect and repair, and the plaintiff, a passer-by on the highway, was injured by falling down a light shaft, whose cover was defective, the landlord was held liable,[1] although the third party could have elected to sue the

---

[1] *Heap* v *Ind Coope & Allsop Ltd* [1940] 2 KB 476, CA.

tenants or to render them jointly liable with the landlord.[2] In this, the landlord could not avoid liability merely by the fact that he took a covenant to repair from the tenant. The same principle, grounded on business efficacy, extends to landlords' implied rights of entry, inspection and repair, so that where a third party was injured as a result of the collapse of a defective wall abutting the highway, he was held entitled to damages from the landlord rather than the weekly tenant – the landlord having an implied right to enter.[3] The existence of an express, implied, or statutory right of entry in favour of the landlord to enter, inspect and repair suffices to attract a potential liability in the landlord. It is not a condition precedent to the landlord's liability that he is under express, implied or statutorily-imposed repairing covenants as against the tenant.

B – Landlord licences tenant to commit nuisance

Only in exceptional circumstances is it possible to hold a landlord liable in nuisance to an adjoining tenant, because of the use made by a neighbouring tenant of adjoining premises let to him by the same landlord, even if the premises are not built to the latest standards. The complaining tenant must prove that the landlord authorised the neighbouring tenant to commit an actionable nuisance.[4] However, where a landlord let premises which lacked sound insulation to two tenants who merely used the premises for domestic living, the fact that he had not provided sound insulation at the time the premises were built or converted did not render the landlord liable in nuisance, since the ordinary use of residential premises without more by the tenants concerned was not capable of amounting to a nuisance. Hence the landlord could not logically be said to be authorising the committing of a nuisance. Nor was he

---

[2] See *Brew Bros Ltd* v *Snax (Ross) Ltd* [1970] 1 QB 612, CA.
[3] *Mint* v *Good* [1951] 1 KB 517, CA.
[4] See *Sampson* v *Hodson-Pressinger* [1982] 1 EGLR 50, CA, where the landlords, having demised a ground-floor flat to the plaintiff, then converted or adapted a roof terrace in a flat above that of the plaintiff tenant, so that this altered terrace could not be reasonably used by the tenants above the plaintiff except in such a way as to cause a nuisance by noise to the latter. Thus the use by the tenant of the upper flat in that case was not showing, in Lord Hoffmann's words, "reasonable consideration" for the occupier of the flat beneath (*Southwark London Borough Council* v *Mills* [1999] 3 EGLR 35 at 40A).

under any "positive duty" in Lord Millett's words, to soundproof the walls between the adjoining premises to keep the noise in.[5]

C – Premises abutting the highway

In the case of a public nuisance caused by a dangerous or defective state of premises abutting the highway, a landlord or owner may be strictly liable in nuisance,[6] to any third party on principles to be discussed. This rule is exceptional.[7] It may be based on the need to protect the safety of the public passing by on the highway.[8] Apart from the case of premises abutting the highway, only if there has been a neglect to repair or to inspect when circumstances called for action on the part of the defendant, and damage or injury to a neighbouring third party is reasonably forseeable as a result of his conduct, is the defendant liable. Where personal injuries result from dangerous or defective premises abutting the highway, the third party may elect to bring an action against any tenant, but the landlord will not be able to avoid liability simply by taking a covenant to repair from the tenant.[9]

Under the strict liability principle, if, owing to a want of repair, premises on a highway become dangerous and constitute a nuisance, so that they collapse and cause injury to a passer-by or to an adjoining owner's premises, the owner or occupier of the dangerous premises, if the latter has undertaken a duty to repair

---

[5] *Southwark London Borough Council* v *Mills, supra* at 42F. Soundproofing is required for buildings erected after, but not before, the Building Regulations 1985, SI 1985 No 1065, and 1991, SI 1991 No 2768, came into force.

[6] Despite the fact that it is established that ordinarily, the law of nuisance is based on reasonable user – the principle of give and take (per Lord Goff in *Cambridge Water Co* v *Eastern Counties Leather plc* [1994] 2 AC 264 at 299); also *Southwark London Borough Council* v *Mills, supra*. In neither case were the specific cases dealing with premises abutting the highway expressly overruled. They cannot be considered to have been impliedly overruled, as policy favours a strict rule in this one instance.

[7] It has been rejected by the Nova Scotia Supreme Court in *O'Leary* v *Meiltides and Eastern Trust Co* (1960) 20 DLR (2d) 258, esp pp 266–268, as being inconsistent with *Sedleigh-Denfield* v *O'Callaghan* [1940] AC 880, as well as on principle (as to which see also (1940) 56 LQR 140).

[8] See Winfield (1940) 56 LQR 1.

[9] *Mint* v *Good* [1951] 1 KB 517, CA.

them, is liable in nuisance. This rule applies whether or not the defendant knew or ought to have known of the danger.[10] The owner's obligation to repair and associated rights of entry to the premises concerned are taken to give him a sufficient degree of control for the purposes of liability. One critic[11] noted that on the facts of the case on which this proposition is based, the premises did not abut the highway: the house concerned fell into neighbouring premises, and there seemed little doubt that the defendant, who had become aware of the defective condition of his house, had neglected to put matters right in time to prevent a collapse. The formulation of a strict liability test was thus not necessary on the facts of that case.

Strict liability extends to any case where, although not under an express duty to repair, the landlord is under an implied duty to do so. It also applies where he has expressly or impliedly reserved the right to enter and inspect and carry out repairs, or even where the landlord has been given permission to enter and repair.[12] Moreover, if statute reserves a right of entry to repair, then the principle will render the landlord liable in nuisance for dangerous or defective premises abutting the highway.

The present principle has some exceptions.[13] If the condition of the premises is caused by the act of a trespasser, or by some cause other than neglect to repair, such as a secret and unobservable operation of nature such as a subsidence under or near the foundations of the premises, a landlord or owner will only be liable in nuisance to third parties if he adopts or continues the nuisance, that is, if he is guilty of negligence. Thus, on ordinary principles, if a third party wishes to rely on continuation of a latent defect and so of a nuisance as a ground of liability, he must show that the defendant had actual or presumed knowledge of the defect and that he, with that knowledge, failed to take any reasonable means to end the nuisance, though with ample time to do so. If adoption of a latent defect is sought to be relied on, the third party must show that the defendant made use of any thing or matter constituting the nuisance.[14] A

---

[10] *Wringe v Cohen* [1940] 1 KB 229, CA.
[11] Friedmann (1939) 3 MLR 305.
[12] *Mint v Good, supra.*
[13] Per Atkinson J in *Wringe v Cohen, supra* at 233.
[14] *Sedleigh-Denfield v O'Callaghan* [1940] AC 880 at 894 (Viscount Maugham); also 913 (Lord Romer). For an example of failure to abate, see eg *Slater v Worthington's Cash Stores (1930) Ltd* [1941] 1 KB 488.

landlord has a reasonable time as from knowledge, or means of knowledge, in which to execute the necessary repairs to make the premises safe: what is a reasonable time is a question of fact and will no doubt vary with the urgency of the repairs.[15] If a third party at the time of the injury caused by the defect in the demised premises has deliberately, rather than accidentally, deviated from the highway, the strict rule does not apply, showing the exceptional nature of the principle.[16]

D – Landlord lets premises with a nuisance

Where a landlord lets premises which are, at the date of the lease, in such a dangerous state as to amount to a nuisance, and he created or continued the nuisance, then he is liable to a third party injured, or whose adjoining premises are damaged, as a result of the condition of the landlord's premises.[17] In one case, the landlord let a building whose chimneys were in a dilapidated condition. The chimneys, which were, to the landlord's knowledge, insecure at the date of the lease, fell onto and damaged the roof and other parts of the plaintiff's adjoining building, and the landlord was held liable in nuisance.[18] A landlord may be liable in nuisance if he knows, or is presumed to have known, of the dangerous or dilapidated condition of the premises at the date of the letting, and so must be held responsible for the continuance of the nuisance resulting from that condition – the basis of his liability being therefore in negligence. If the landlord is liable for repairs, as against the tenant, this in itself suggests that he has continued the nuisance.[19]

Even if the landlord is not subject to any covenant to repair, he may still be liable to a third party for damage caused to him by a patent or a latent defect in the demised premises, if he has sufficient actual or presumed knowledge or control of the premises at the date of the letting. Thus, a landlord was liable to a third party where he had let premises to the tenant in a condition which the landlord was presumed to have known to be defective at the date of the lease:

---

[15] *Leanse v Egerton* [1943] KB 323 (owner presumed to have knowledge as from next day after damage by bomb).
[16] *Jacobs v London County Council* [1950] AC 361, HL.
[17] *Todd v Flight* (1860) 9 CB (NS) 377.
[18] *Todd v Flight, supra.*
[19] See *Pretty v Bickmore* (1873) LR 8 CP 401.

there was seepage under a flank wall in the premises, causing the wall to tilt towards the neighbouring premises.[20] Conversely if, at the date of the lease, the landlord has no reason to suppose that there is a nuisance and could not, at the date of the letting, have discovered it with the exercise of reasonable care, then he, but not his lessee, escapes liability. Where part of an old city wall, let to a weekly tenant, fell into and damaged the adjoining premises of a third party, since the landlords had no knowledge or means of knowledge of the wall's condition, they were, accordingly, held not liable.[21]

If the landlord is liable in nuisance for letting premises in a defective state, he cannot avoid liability by taking from the tenant a covenant to repair; also, if the landlord is liable for repairs and has a right of entry, express, implied, or under statute, to inspect and carry out repairs, then he will be liable in nuisance for defects which he ought to have discovered, on principles already discussed.

III – Liability of lessees in nuisance

A – General

A lessee may be liable to third parties in nuisance occasioned by the defective or dangerous state of premises.[22] Apart from public nuisance, viz, where the premises adjoin the highway, liability is based on a neglect to repair. Where, therefore, a lessee under a full repairing covenant held a lease of a building and the plaintiff was a monthly subtenant of part of the premises, and a heavy piece of guttering fell through the roof of that part and injured the plaintiff's wife, he was held entitled to damages from the head lessee, who was under a duty, which he had failed to observe, to take care to prevent the guttering from falling.[23] If a person is responsible for continuing or adopting a nuisance, as where a lessee is under a liability to keep premises adjoining the highway in repair and neglects to do so, he will be strictly liable to third-party highway users and adjoining owners for damage or injury caused by defects in the premises.[24]

---

[20] *Brew Bros Ltd v Snax (Ross) Ltd* [1970] 1 QB 612, CA.
[21] *St Anne's Well Brewery Co v Roberts* (1928) 44 TLR 703, CA.
[22] *Cheetham v Hampson* (1791) 4 TR 318.
[23] *Cunard v Antifyre Ltd* [1933] 1 KB 551. It made no difference to the result that the two tenements were part of the same structure.
[24] *Pemberton v Bright* [1960] 1 WLR 436.

B – Public nuisance

A lessee is under a duty to any users of the highway to see to it that no injury results to them from any defective condition of the demised premises. This particular duty has been formulated in terms of strict liability. Where, therefore, a passer-by on the highway was injured by a falling heavy lamp in front of a house, the third party recovered in nuisance, without any proof of negligence by the occupier, a lessee of the premises. It was assumed that the latter would regularly inspect the lamp.[25] The liability of any lessee in relation to premises abutting the highway is co-extensive with any liability of the landlord's. Oral tenants were held jointly liable with their landlord to a third party for injuries caused by the fall of a defective shutter. The landlord had reserved himself a right to enter and repair, even though he was subject to no obligation to repair.[26]

An occupying lessee's strict liability in public nuisance in relation to premises abutting the highway is subject to some few exceptions, analagous to those applying to landlords. Thus, if the lessee has no knowledge or means of knowledge of the nuisance, capable of being discovered by reasonable inspection, as where it is a secret, latent defect, or if he does not continue the nuisance with such knowledge, he will not be liable.[27] Regular inspections should reduce the risk of liability.

If the third party is not a user of the highway at the time of the injury, but instead is a lawful visitor to the premises, then if the third party is injured by a defective part of the premises, such as a cornice, falling on him, the lessee will only be liable, first, if the defect was patent as opposed to latent, and, secondly, if the third party is able to prove that the defendant knew or ought to have known of the defect.[28] In other words, the strict liability rules do not extend beyond the protection of highway users.

---

[25] *Tarry v Ashton* (1876) 1 QBD 314.
[26] *Wilchick v Marks and Silverstone* [1934] 2 KB 56. The tenants had to bear the whole loss despite their issuing third party notices against the landlords because they could not prove a contract by the latter to do any repairs.
[27] *Barker v Herbert* [1911] 2 KB 633, CA; *Noble v Harrison* [1926] 2 KB 332.
[28] *Pritchard v Peto* [1917] 2 KB 173 (where the case was pleaded in negligence).

IV – Liability of occupiers to lawful visitors and others

A – *Scope of liability*

The Occupiers' Liability Act 1957 imposes on any "occupier" of premises liability towards any lawful visitor where he is injured by a defect on the premises. It appears that "occupier" is "a convenient word to denote a person who had a sufficient degree of control over premises to put him under a duty of care towards those who came lawfully onto the premises".[29]

While in the case of an occupying freeholder in possession and control of the whole of premises, there is no difficulty in treating him as the "occupier", different considerations arise where the whole or part of premises have been let, in which case it seems that the lessee would be an "occupier", even, arguably, if he has granted a licence to occupy the whole or some part of the premises to a licensee, seeing that ordinarily a lessee does not give up exclusive possession to his licensee.[30]

Two persons may be "occupiers" of the same premises, as where landlords remained the "occupiers" of living accommodation used by their manager and the latter's paying guests, seeing that the manager was their licensee.[31] Indeed, where two persons are "occupiers" of the same premises, a different level of the common duty of care may be imposed on each occupier, so that the owners of a public house were held to be liable to see that in the private portion of a public house, the structure was reasonably safe, but not that lights had been properly switched on and rugs safely laid on the floor.[32] Different persons may be "occupiers" of different parts of a building, as where a landlord lets the whole of a house minus some specified or excepted part. If, for example, a staircase or a common roof has been reserved from a lease, the landlord

---

[29] *Wheat v E Lacon & Co Ltd* [1966] AC 552 at 577F (Lord Denning).
[30] At least where the licence had been deliberately granted so as to avoid the 1957 Act and possibly in any case: see *Kelly v Woolworth & Co* [1922] 2 IR 5; *Salmond and Heuston on the Law of Torts*, p 260.
[31] *Wheat v E Lacon & Co Ltd, supra*; or where an owner let a flat to service occupiers who allowed their child to play on an abutting roof: *Bailey v Armes* [1999] EGCS 21, CA. However, in the latter case the service occupiers were held not to satisfy the control test.
[32] *Wheat v E Lacon & Co Ltd, supra* at 581 (Lord Denning); also at 585–586 (Lord Morris).

seemingly remains the "occupier" for statutory purposes of the excepted parts of the premises.[33]

B – Common duty of care

The Occupiers' Liability Act 1957 imposes on any occupier a "common duty of care" to any lawful visitor, in respect of the whole premises, if he is there invited, or to the part of them to which he is invited. "Lawful visitor" includes any person lawfully entering the premises, as opposed to a trespasser. This expression also includes, by section 2(6), persons who enter the premises in the exercise of a right conferred by law, whether they have the occupier's permission or not. Under the general law, the complainant who is injured must also prove that injury of a given description was reasonably foreseeable, although he does not have to prove that the precise injury was reasonably foreseeable.[34]

By section 2(2) of the 1957 Act, the common duty of care is a duty to take such care as, in all the circumstances of the case, is reasonable to see that the visitor will be reasonably safe in using the premises for the purposes for which he is there. The duty applies to protect the lawful visitor's person and property. It is "not to ensure the entrant's safety, but only to show reasonableness".[35] The duty is thus subjective, and varies with the circumstances. It is likely to be higher in the case of children and elderly persons and relatively lower in the case of a normally active adult or adolescent.[36]

In the words of Lord Steyn, "in this corner of the law the results of decided cases are inevitably very fact-sensitive". To him, the rest of the House of Lords agreeing, comparing the facts of one case with another was a "sterile exercise", at least where it led to no discernable principle being arrived at.[37] These observations justify only a selective use of illustrations. Where a gap between the threshold of the doorway to a security kiosk and the ground was too wide to be safe, and the plaintiff injured his foot when entering

---

[33] *Wheat* v *E Lacon & Co Ltd*, supra at 579 (Lord Denning).
[34] *Jolley* v *Sutton London Borough Council* [2000] 1 WLR 1082. Lord Hoffmann cited at 1091E the example of it being foreseeable that a child falling through a hole in the street would be injured.
[35] *M'Glone* v *British Railways Board* 1966 SC (HL) 1 at 15 (Lord Guest).
[36] *McGinley* v *British Railways Board* [1983] 1 WLR 1427, HL.
[37] *Jolley* v *Sutton London Borough Council* [2000] 1 WLR 1082 at 1089E–F.

the kiosk, the occupier broke his statutory duty and was answerable in damages.[38] Similarly, a local authority occupier's failure to replace glass in a school door – the thickness of the glass being no longer in conformity with the latest British Standards – amounted to a breach of statutory duty.[39] Many of the cases have been concerned with dangerous defects on the premises, but in each case the extent of an occupier's duty will vary with such things as the type of premises[40] and the activity of the visitor, as well as the part of the premises being used. Thus, where a pathway regularly used for access to their premises by old people became irregular, so that a visitor was injured after a fall on a projecting piece of paving stone, a local council was in breach of its statutory duty in failing to make the path safe for this type of visitor, who would be likely to be unsteadied by variations in the path with potentially serious consequences.[41] If a visitor is on the premises as the result of a contract, much depends on the nature and incidents of the contract. If, under the contract, the occupier is bound to allow third parties to enter and use the premises (such as allowing onto them a builder's workmen or the occupier's own subcontractors), section 3 of the 1957 Act will apply. Section 3(1) states that where the occupier is bound to permit strangers to a contract to enter or use the premises, his duty of care cannot be restricted or excluded by the contract. If the contract imposes a higher duty, then the stranger is entitled, by section 3(1), to the benefit of that duty. By section 3(2), unless the contract expressly so provides, the occupier will not be liable to strangers for any injury caused by faulty work other than faulty work done by himself, his servants and persons acting under his direction or control.

C – *Special factors*

1. Section 2(5) of the 1957 Act provides that the common duty of care does not impose on the occupier any obligation to a visitor in respect of risks voluntarily accepted by him as visitor.

---

[38] *White v Lord Chancellors Department* [1997] CLY 3805.
[39] *J (A minor) v Staffordshire County Council* [1997] CLY 3783.
[40] Thus a landlord was not obliged to keep a particular common stairway in a block of flats lighted at all hours of darkness: *Irving v London County Council* (1965) 109 SJ 157.
[41] *Wright v Greenwich London Borough Council* [1996] CLY 4474.

2. Circumstances relevant to the common duty of care include, by section 2(3), the degree of care, and want of care, which would ordinarily be looked for in a visitor. For example, "when a householder calls in a specialist to deal with a defective installation on his premises, he can reasonably expect the specialist to appreciate and guard against the dangers arising from the defect."[42] Thus, if a visitor on the premises is there in the exercise of his calling, he will appreciate and guard against any special risks ordinarily incident to it, in so far as the occupier leaves him free to do so (s 2(3)(b)).[43]
3. Section 2(3)(a) of the 1957 Act provides that an occupier must be prepared for children to be less careful than adults. He might, therefore, have to warn a child of a danger which would be obvious to an adult, as with an unlit passageway or staircase. In addition, the occupier may now have to remove or cure any obvious or even less apparent source of danger on the premises, having regard to the ingenuity of at least unsupervised children to find unexpected ways of doing mischief to themselves.[44] But because the Act does not treat an occupier as an all-risks insurer, he is probably still not bound to assume that a child's parents will not, in relation to an open space, which might contain obvious hazards to any parent, fail to be prudent and so exercise due care in supervising the child's activities in playing there.[45]

D – Subjectivity of common duty of care

Section 2(4) of the 1957 Act provides that in determining whether the occupier has discharged the common duty of care, regard is to be had to all the circumstances. This provision was enacted because it had been held, prior to the 1957 Act, that if a person came onto the premises as an invitee, and had been injured by the defective

---

[42] *Roles v Nathan* [1963] 1 WLR 1117 at 1123 (Lord Denning MR).
[43] For a case where a visitor (an experienced surveyor) had his damages reduced by one-third because he ought to have realised the hazardous nature of the darkened room he was visiting, see *Rae v Mars (UK) Ltd* [1990] 1 EGLR 161.
[44] *Jolley v Sutton London Borough Council* [2000] 1 WLR 1082 (disused boat with warning notice on it used normally for abandoned cars); see J Murdoch "Don't Play Around" *Estates Gazette*, August 19 2000 p 73.
[45] *Phipps v Rochester Corporation* [1955] 1 QB 450.

state or condition of the premises, due to the occupier's fault, the occupier had a complete defence if he proved that the visitor knew of the danger or had been warned of it.[46] Relevant factors might include the type of premises, their standard of lighting, the age and purpose of the visitor, the nature of the defect causing the injury and the visitor's state of knowledge. The question of whether any warning of the danger was given by notice, or otherwise, is material, but not, having regard to the intent to modify the previous common law rule, decisive. Hence, section 2(4)(a) provides that where damage is caused to a visitor by a danger of which he had been warned by the occupier, the warning is not, by itself, conclusive to absolve the occupier from liability, unless it was enough to make the visitor reasonably safe. If a warning notice would not have made any difference to the visitor's conduct, it seems that he cannot reasonably complain of the absence of any notice.[47]

If a visitor suffers personal injury or damage to his property from a danger caused by faulty execution of work of construction, maintenance or repair by an independent contractor employed by the occupier, the latter is, by section 2(4)(b), absolved from liability if, in all the circumstances, he acted reasonably in entrusting the work to an independent contractor and if he took all reasonable steps to satisfy himself that the contractor was competent and the work was properly done. The House of Lords have held that the purpose of this provision is to "afford some protection from liability to an occupier who has engaged an independent contractor who has executed the work in a faulty manner".[48]

The conduct of the visitor may be wholly or partly responsible for his injuries, and this may be taken into account to eliminate or reduce the occupier's liability.[49] The courts are reluctant (in the absence of special factors such as a known practice on the part of

---

[46] *Roles* v *Nathan, supra* at 1124 (Lord Denning MR).
[47] *Staples* v *West Dorset District Council* (1995) 93 LGR 536.
[48] *Ferguson* v *Welsh* [1987] 1 WLR 1553 at 1560 (Lord Fraser). Despite the use of the expression "the work had been properly done" in s 2(4), it applies both where work has been completed and also to those where it is in progress. If the occupier had reasonable cause to suspect that the independent contractor was using an unsafe system might it be reasonable to expect the occupier to exercise supervision over him: *ibid*.
[49] See eg *Whyte* v *Redland Aggregates Ltd* [1998] CLY 3989.

visitors to undertake a given activity[50]) to fix an occupier with liability under the common duty of care where the injuries to the visitor result from his own folly – as where a plaintiff fell on a wall slope which any sensible person would have made no use of, the dangers of doing so being all too obvious.[51]

E – Exclusion of common duty of care

Section 2(1) of the 1957 Act preserves the freedom of an occupier to restrict, modify or exclude his duty to visitors by agreement or otherwise. This right could be achieved by a suitably prominent notice at the entrance of the occupier's land or premises, which makes it clear that entry is conditional on visitors doing so at their own risk.[52] It makes no difference that the visitor does not see or read the notice.

F – Limit on right to exclude liability

Where an occupier retains part of the demised premises in his control, section 2(1) of the Unfair Contract Terms Act 1977 provides that he cannot exclude or restrict his liability for death or personal injury resulting from negligence, by any notice. In relation to other loss or damage, an occupier will only be able to exclude or restrict his liability for negligence if the notice satisfies the statutory requirements of reasonableness (section 2(2)).

These restrictions on the freedom of an occupier to exclude liability apply only to business liability (s 1(3) of the 1977 Act). It is not clear whether a narrow interpretation applies to the expression "business", on the ground that the legislation is an interference with freedom of contract, or a wide view of the expression, on the ground that the 1977 Act amendments were for the protection of the public.[53] A domestic occupier remains free to restrict, modify or exclude his liability for negligence by contract or notice. But if a domestic occupier allowed part of his house to be occupied by a lodger, the restrictions in the Unfair Contract Terms Act 1977 might

---

[50] As in *Harrison* v *Thanet District Council* [1998] CLY 3918.
[51] *Staples* v *West Dorset District Council* (1995) 93 LGR 536.
[52] *Ashdown* v *Samuel Williams & Sons Ltd* [1957] 1 QB 409, CA.
[53] Mesher, "Occupiers, Trespassers and the Unfair Contract Terms Act 1977" [1979] Conv 58, 58–60.

protect the lodger since taking in a lodger could be said to be a business activity.[54]

The limitations on business occupiers' powers to contract out of their common duty of care do not apply where such an occupier permits access to his premises for recreational or educational purposes.[55] The policy of the exemption is evidently to see to it that farmers, for example, who allow access to their land for recreational or other purposes are not put off from allowing that access by the possibility of legal liability.

G – Duty of occupier to "trespassers"

The occupier's duty to "trespassers", or rather in the statutory language, to "persons other than visitors", is intended to be less extensive than his duty to lawful visitors.[56] A person entering premises with a criminal intent, such as a burglar, is not entitled to the same degree of protection against injury from a dangerous defect there than would be a lawful visitor.

In relation to the risk of personal injuries to "trespassers"[57] on the premises, section 1(3) of the Occupiers' Liability Act 1984, which revised the law, provides that the occupier owes a duty to a "trespasser" if:

(a) he is aware of the danger (on the premises) or has reasonable grounds to believe that it exists;
(b) he knows or has reasonable grounds to believe that the other is in the vicinity of the danger or that he may come into its vicinity;[58]

---

[54] See Mesher, *op cit, supra*.
[55] However, those business occupiers providing such facilities as a school or a university remain subject to the 1977 Act limitations.
[56] See Buckley "The Occupiers' Liability Act 1984" [1984] Conv 412.
[57] The courts still favour this expression: see eg *Ratcliff v McConnell* [1999] 1 WLR 670 (where a student who dived head first into a college swimming pool, knowing it to be closed for the winter, failed, knowing the risks, to recover damages).
[58] ie actual knowledge or reasonable grounds for a belief by the occupier that trespassers would be present (as from part experience): *Swain v Natui Ram Puri* [1996] CLY 5697, CA. It was there said that it is not sufficient for the claimant to prove constructive knowledge: the legislation omits the words "ought to have known".

(c) the risk is one against which in all the circumstances of the case he may reasonably be expected to offer some protection to the trespasser.

There is some uncertainty as to the borderline between a trespasser and a lawful visitor. According to a leading work,[59] no person is a trespasser (or a "person other than a visitor") if he enters the premises with a view to any kind of communication with the occupier or other person on the premises, whether he knows or not that his entry is prohibited. It is not clear whether such persons as canvassers, who may not be welcome on the premises, as well as, say, those who visit a shop or other trade premises not with a view to purchasing or window shopping but with the intention of being competitive observers,[60] would be lawful visitors or "trespassers". The former would seem to be lawful visitors unless and until the occupier intimates that they are not welcome. The latter, by contrast, would seem by the nature of their purposes on the premises to be trespassers *ab initio*, but the point remains to be settled in this jurisdiction.

The new statutory test is a general formulation only and now seems "not [to] represent a significant advance upon the supposedly uncertain and unpredictable principles of 'common humanity'".[61] At all events, the duty of the occupier, by section 1(4), is to take such care as is reasonable in all the circumstances of the case to see that the "persons other than visitors" do not suffer injury on the premises by reason of the danger concerned. The duty may, by section 1(5), be discharged in appropriate cases by taking such steps as are reasonable in all the circumstances to give warning of the danger or to discourage persons from incurring the risk. The doctrine of *volenti non fit injuria* or voluntary assumption of risk is expressly preserved (s 1(6)). If a warning notice is put up at the entrance to or boundary of the land or premises, and it is sufficiently prominent and legible to be capable of being read by

---

[59] *Salmond and Heuston on the Law of Torts*, 21st edn (1996), p 277, citing *Christian* v *Johanssen* [1956] NZLR 664.
[60] See *Chaytor* v *London, New York and Paris Association of Fashion and Price* (1961) 30 DLR (2d) 527.
[61] Buckley, *supra*, p 418. Indeed, in *Ratcliff* v *McConnell* [1999] 1 WLR 670 it was held that the pre-1984 test of liability for trespassers was still the basis of the law.

any trespasser, the occupier will probably be able to avoid liability.[62] This is owing to the fact that if a "person other than a visitor" ignores a warning notice, he enters *volens*. Warning notices might not suffice to avoid liability in the case of child trespassers. No notices cast in the form "trespassers will be prosecuted" could, seemingly, avoid liability in relation to concealed or unusual hazards on the land or premises in respect of any trespasser, with the somewhat odd result that a "person other than a visitor" is to that extent placed in a better position than a lawful visitor.

V – Access to neighbouring land

A – Introduction

The Law Commission, noting that the common law provides no general right of access to enable a property owner to gain access to neighbouring land to carry out repairing or other maintenance work, if a neighbouring owner refuses such access, recommended reforms.[63] The fact that a person could procure access to neighbouring land or premises for repairs by means of a licence was admitted. A licence might not be conferred, would be personal to the parties and would probably have to be paid for. The Commission were concerned at the risk of deterioration of property if the owner could not obtain access to neighbouring land for essential work.

The Access to Neighbouring Land Act 1992 attempts to balance the possibly oppressive conferral of a general right of access by a number of safeguards to the "respondent" with respect to the "servient land", notably respecting the terms of any access. The fact that access is only allowed under a county court order was seemingly thought to provide sufficient safeguards for both parties. The Act thus provides a "partial solution"[64] to the problem arising as where, for example, an owner needs access to neighbouring land in order to repair a building which has been declared dangerous by a local authority.

The 1992 Act is less radical in terms than the Queensland Property Act 1974.[65] Section 180(1) of that Act imposes a "statutory

---

[62] *Ashdown v Samuel Williams & Sons Ltd, supra.*
[63] Law Com No 151 (1985) *Rights of Access to Neighbouring Land.*
[64] Megarry and Wade, 18–224. See also Cheshire and Burn p 583; also *John Trenberth Ltd v National Westminster Bank Ltd* [1980] 1 EGLR 102, noted Street [1980] Conv 308.
[65] See Tarlo "Forcing the Creation of Easements – a Novel View" (1979) 53 Australian Law J 254.

right of user" in respect of servient land, which the servient owner is bound to permit, if it is ordered, on the broad ground of reasonable necessity in the interests of the "effective use in any reasonable manner" of the dominant land. This right of user (which may be in the form of an easement, licence or otherwise and may be perpetual or for a fixed period) is triggered only if the servient owner is found, by the court, unreasonably to have refused the imposition of a right of user over his land (s 180(3)(c)).[66] There is a wide public interest requirement – it must be consistent with the public interest that the land should be used in the manner proposed (s 180(3)(a)). Adequate money compensation must be paid, in principle, to the owner of the servient land for any loss or disadvantage suffered from the imposition of a right of user (s 180(2)(b)). The Queensland legislation allows the court more flexibility than the Access to Neighbouring Land Act 1992, with its careful system of detailed checks and balances. The nature and duration of the right to be imposed by the court on the servient owner may go further than in England, where the creation of a permanent easement is not envisaged, or indeed the creation of any right in property. It also applies well beyond specified types of work, so sparing some of the definitional uncertainties which may go with the approach of the 1992 Act. At the same time, there is a danger of a wider right of access than that conferred in the 1992 Act being seen as an upset to property rights and as encouraging disputes.[67] For this reason, it is not clear that the broad approach in vogue in Queensland has any more to commend it than the narrower basis of our own regime.

B – General effect of Access to Neighbouring Land Act 1992

The 1992 Act allows any "person" to gain access to neighbouring land for the purpose of carrying out repairs and related work on that land which are required for the benefit of his own land. This

---

[66] As where the owner demands an unconscionable amount of compensation: *Re Seaforth Land Sales (No 2)* [1977] QdR 317, where the court ordered a reduction in the sum demanded. A similar principle might apply where entry is refused, as envisaged by s 1 of the 1992 Act.

[67] See PH Kenny's notes to the 1992 Act in Current Law Statutes Annotated; also Barsby 150 (2000) New Law J 1256.

seemingly personal right,[68] characterised as a "short-term limited-purpose easement of access to a landowner over a neighbour's land"[69] is dependent on and subject to the terms of an access order, obtained in the county court, which, in its wide discretion, fixes the terms, conditions and duration of the access. An applicant cannot normally ask for an access order to carry out any improvements or alterations to his land. It remains possible for neighbouring owners to agree upon an access to the relevant land by a licence. What neither means of access can now do is to preclude an application to the court, as the right to apply for an access order cannot be contracted out of or restricted (s 4(4)). The uncertainties of the Act are not assisted by the statutory references to "reasonably necessary", as the criterion for an access order; the fact that that the "respondent" to the order is not specified in the Act; and that the flexible term "basic preservation works" is used to describe the works within the Act.

The general result of the 1992 Act is to enable a person (not neccesarily the owner of the land) to obtain an order under which he gains entry, immune from an action for trespass, to his neighbour's land,[70] to undertake "basic preservation works" to his land, notably in a case where the work would otherwise be impossible, or the expense of doing the work by means of access is much less than it would be if the work were done wholly from the person's own land. If the court so wishes, it may discharge or vary the terms of any access order (s 6(1)). Should any person (eg the respondent or his lessee) fail to comply with any term of an access order, he is liable to an action for breach of statutory duty which sounds in damages.

C – Access orders

A person who for the purpose of carrying out works to any land (called the "dominant land") desires to enter upon any

---

[68] Megarry and Wade, 18–227.
[69] HW Wilkinson "Please May We Have Our Ball Back?" [1992] Conv 225 at 230.
[70] The expression "land" is not defined in the 1992 Act but it was held wide enough to include a party wall under the pre-Party Wall etc Act 1996 rules prevailing outside Inner London: *Dean v Walker* (1996) 73 P&CR 366, CA.

neighbouring or adjacent land, and who needs, but does not have, the consent of some other person to the entry, may apply to the court for an access order (s 1(1)). Applications are to be commenced in the county court (s 7(2)). Since an access order may only be applied for by a person desirous of carrying out "basic preservation works", the common law will govern any access for works of improvement or alteration, sometimes thus requiring the licence of the "servient owner", if there is no right of access pursuant to an express or implied easement.

The applicant may be the freeholder of the dominant land, or any person with an estate or interest in that land, whether as long lessee, periodic tenant or mortgagee; but technically applicants could include a licensee of the land and any building contractor. The respondent is normally the owner and any person in actual occupation at the relevant time; this might be the freeholder, or, if the latter is not in occupation, any lessee or licensee. The Act seems to ignore the risk that it may be a breach of covenant for a lessee to allow third parties access to the premises. Whether the lessee would in such a case have a defence to a lessor's action for breach of covenant is not clear.

D – Basic preservation works

Once an application is made, the county court has no automatic power to make an access order. The court must be satisfied (s 1(2)) that:

(a) the works are "reasonably necessary for the preservation of the whole or any part of the dominant land"; and
(b) the works cannot be carried out, or would be "substantially more difficult to carry out" without entry to the servient land.

If the applicant satisfies the court that it is reasonably necessary to carry out "basic preservation works" to the dominant land, he is presumed to satisfy the requirement that the works are reasonably necessary for the preservation of the land (s 1(4)). Thus, if a roof leak may only be mended by access from next door, there is a presumption that such work is necessary in the required sense. The same would arguably apply to a flank wall not accessible from the dominant premises.

Because it is the policy of the Act to exclude work of improvement or alteration, the expression "basic preservation works" is defined, seemingly exclusively, (s 1(4)) as meaning any of four matters:

(a) "the maintenance, repair or renewal of any part of a building or other structure comprised in, or situate on, the dominant land";
(b) "the clearance, repair or renewal of any drain, sewer, pipe or cable in or on the dominant land";
(c) the treatment, cutting back, felling, removal or replacement of any hedge, tree, shrub or growing thing on the dominant land;[71]
(d) "the filling in or clearance of any ditch" on the dominant land.

E – Statutory defence

The respondent to the application, as owner or occupier of the servient land, has one of two defences which, if made out, preclude the making of an access order. The intention of these defences may be to discourage trivial applications.

The first defence is that the respondent or any other person (such as his lessee) would "suffer interference with, or disturbance of, his use or enjoyment of the servient land" to such a degree, by reason of the entry, that it would be unreasonable to make the order (s 1(3)(a)). The second defence requires proof by the respondent that he, or any other person, in occupation of the whole or any part of the servient land, would suffer hardship (s 1(3)(b)) – again to such a degree that it would be unreasonable to make an order. Since some interference or hardship is inevitably going to result from the making of an access order, the respondent must show that the degree of interference or hardship makes it unreasonable to make the order, and that any terms of the order would not safeguard the respondent.

F – Further analysis of the provisions

The court may qualify the essential principle of the 1992 Act to confine access orders to repairs and maintenance work, if it considers such a decision to be "fair and reasonable in all the circumstances of the case" (s 1(5)). The court may then treat certain works, for which access is sought, as being "basic preservation

---

[71] In this case the item concerned must be or be in danger of becoming damaged, diseased, dangerous, insecurely rooted or dead.

works" notwithstanding the fact that the works "incidentally involve":

(a) the making of some "alteration, adjustment or improvement" to the land; or
(b) the "demolition of the whole or of any part of a building or structure comprised in or situated on the land".

The difficulty is to decide when works of, say, improvement, as opposed to repair, are "incidental" in the required sense, otherwise the exception destroys the generality of section 1(4) of the Act. Section 1(5) seemingly allows, for example, the court to permit an applicant to include work of replacing items which are not out of repair because they are not damaged (such as wooden or metal windows or doors) with differently designed items, a process which at common law is not a repair. The scope of the second extension of section 1(5) is obscure, seeing that demolitions incidental to "basic preservation work" are already allowed by the reference in section 1(4)(a) to "renewal", but the Law Commission justified this extra extension partly in respect of the removal (seemingly without rebuilding) of unsafe buildings.[72]

G – Terms and conditions of access orders

An access order must specify the works in question (s 2(1)(a)) and the particular area of the servient land to be entered for the purpose of the works (s 2(1)(b)) and also the date and period of entry (s 2((1)(c)). If the area of the land specified later turns out not to be enough for the applicant's purpose, he must apply to the court for a variation of the order, rather than apply to the person in occupation of the servient land for permission. Sometimes it happens that the work may take longer than the order envisages, or involve further works, than set out in the order: if so, an application to vary is required.

The court has a wide discretion to impose terms and conditions, which it considers necessary to avoid or restrict any loss, damage or injury that might otherwise be caused to the respondent or any other person by the entry, as well as avoiding or restricting any inconvenience or loss of privacy that might otherwise be so caused

---

[72] *Report, supra*, para 4.11.

(s 2(2)). Thus terms and conditions may notably include provisions for the manner in which the specified works are to be carried out (s 2(3)(a)). An access order may require the applicant to pay to the respondent money compensation for the privilege of entering the servient land.[73] The amount is what the court thinks fair and reasonable, having regard to two matters specified in the Act in particular,[74] but not exclusively.

H – Effect of access order

The precise terms of any access order are for the court. The respondent and any successor in title of his to the land are bound by any access order, provided in the latter case that it is registered (s 5).[75] So, subject to the same condition, is any person with an estate, interest or right in the servient land or any part whose interests were created after the making of the order, and who derives title under the respondent, such as lessees and even licensees in respect of the part of the servient land they occupy (s 4(1)). Where an access order is made against joint tenants of servient land, since there is unity of title, all defendants are bound by the order. But where the freehold to the servient land is severed, as where a roof or wall crosses two plots, the order may seemingly only bind the owners to the extent of their interest.[76] As a result of section 4(1)(b), if an access order is obtained against, say, a yearly tenant of the whole servient land, it will bind a landlord by assignment of that land after the making of the order, but not the person who was landlord at the time of the order, who would need to be joined as a party to the application. The respondent is required "so far as he has power to do so" to permit the applicant and "any of his associates" to enter the land and then do anything

---

[73] But not in the case of works on residential land (defined s 2(7)).
[74] ie the "likely financial advantage" to the applicant and persons connected with him such as the landowner and the degree of inconvenience to the respondent and persons connected with him (again such as the owner) resulting from the order in each case (s 2(5)).
[75] The requirement of registration in the case of registered land is satisfied by registration of a notice or caution under the Land Registration Act 1925 s 49 or 54: see Ruoff and Roper, *Registered Conveyancing*, para 35–32A.
[76] Cf PH Kenny, *op cit*, note to s 4(1).

which he or they are authorised to do by the order, for the purpose of the specified works (s 3(1) and (2)(a)) – in relation to the specified land area. Such entry is not a trespass (s 3(6)). An unauthorised entry into land which is not within the specified area in the order would doubtless contitute a trespass.

The applicant and his "associates" (who are widely defined in s 3(7)) may, subject to the court's power to vary or exclude this right, bring on to the land concerned and leave there for the period permitted by the order, and then remove them before the order period expires, "such materials, plant and equipment as are reasonably necessary for the carrying out of the works" (s 3(2)(b)). Once an entry period under an access order ceases, and is not extended, the applicant must leave the land and leave nothing in, on or over the servient land expect so far as required to comply with their duty to make the land good (s 3(3)).

Chapter 14

Miscellaneous aspects

In the following Chapter there are examined a number of diverse aspects of the law of dilapidations.

I – Covenants to repair freehold land

A – General rule

The House of Lords have confirmed a long-standing rule that a covenant to repair freehold land cannot run with the burdened land in equity, so that once the original covenantor has sold his freehold, the covenant cannot be enforced against the purchaser by the neighbouring freeholder.[1] Hence, a claim by a freehold owner of a cottage to damages from his neighbour, for non-repair of a roof which covered both premises, failed. Both owners were successors in title. Lord Templeman said:

> For over 100 years it has been clear and accepted law that equity will enforce negative covenants against freehold land but has no power to enforce positive covenants against successors in title of the land. To enforce a positive covenant would be to enforce a personal obligation against a person who has not covenanted. To enforce negative covenants is only to treat the land as subject to a restriction.[2]

Lord Templeman, speaking for the whole House of Lords, acknowledged the criticisms of this rule, and noted that several reform proposals had been made. In fact, all of these have come to nothing.[3]

---

[1] *Rhone* v *Stephens (Executrix)* [1994] 2 EGLR 181, HL.
[2] *Ibid* at 183.
[3] The general reform package proposed by the Law Commission (Law Com No 127 (1984)) will not, it seems, be implemented. However, if legislation to introduce commonhold flats is introduced (the latest set of proposals being the Commonhold and Long Leasehold Consultation Paper and Draft Bill, published in late August 2000) positive covenants would, within a commonhold scheme, be capable of being mutually enforced by and against unit holders. See also below.

The House of Lords refused to overrule well-established principles applying to freehold land[4] in part because to do so would, in Lord Templeman's words, "create a number of difficulties, anomalies and uncertainties and affect the rights and liabilities of people who have for over 100 years bought and sold land in the knowledge ... that positive covenants affecting freehold land are not directly enforceable except against the original covenantor".[5] Moreover, as noted by Lord Templeman, although positive covenants may be enforced against successors in title to a leasehold interest, Parliament had to legislate to allow enfranchisement so as to prevent the injustices arising out of that rule on the expiry of long leases. Thus, although the rule in relation to freeholds causes some inconvenience,[6] its affirmation by the House of Lords is not obviously incorrect.[7] Unfortunately, the rule renders it largely impossible to enforce covenants to repair and maintain against successors in title to freehold flats, which may explain proposals to introduce commonhold freehold tenure in England and Wales,[8] which, if ever introduced, would make more or less equivalent provision for freehold flats as exists in many other jurisdictions,[9] at the price of provoking a new field for litigants.

Mitigation of the rule

Apart from the granting of a lease of the land concerned, the rule about covenants to repair freehold land may be avoided with some difficulty.[10] Freehold land burdened with certain types of obligation to repair or maintain may be subjected, each time it is conveyed, to

---

[4] As derived from *Austerberry* v *Oldham Corporation* (1885) 29 ChD 750.
[5] *Rhone* v *Stephens, supra* at 183H.
[6] Thus it was extended to an obligation to maintain a hedge in a private Inclosure Act in *Marlton* v *Turner* [1997] CLY 4233.
[7] See the comments of Snape [1994] Conv 477.
[8] As originally proposed by the Aldridge Committee (1987) Cm 179; the latest set of proposals are in a Draft Bill and Consultation Paper of August 2000: see above.
[9] There is a comprehensive review of these in Van der Merwe, "Apartment Ownership", Ch 5 of Vol VI of *International Encyclopedia of Comparative Law*.
[10] See (1993) 137 Sol Jo 938, where the possibility of imposing equitable charges on successors in title to freehold land is canvassed; also Cheshire and Burn pp 668–670.

an indemnity covenant by the latest owner, which chain will be broken if a subsequent covenantor goes bankrupt, and in any case it now appears that any such duty must be linked to a correlative right to benefits. In one case, a sufficient reciprocity of benefit and burden existed, since neither the defendant nor his predecessor in title could expect to benefit from the common use of estate roads and sewers (at least where easements had been granted over them) without contributing to their maintenance; but the defendant could choose not to take the benefits if he did not wish to pay towards their costs.[11] Perhaps the burden in this case was akin to an easement to use a park if one is prepared to pay towards the costs.[12] In the case of fencing in an estate, or moor, or other confined area, an owner cannot now assert a duty to fence against his neighbour without himself undertaking the costs of fencing.

Apart from the doctrine of mutual benefit and burden, any freeholder is entitled to claim an access order under the Access to Neighbouring Land Act 1992 for the purpose of entering neighbouring land to carry out certain types of work (see Chapter 13 of this book), so mitigating any results of not being able to enforce a covenant to repair by his neighbour. Local authorities have powers to control unfit housing which is out of repair (Chapter 10) and the law of nuisance and negligence (Chapter 13) protects third parties against the worst results of disrepair, at least in so far as they impact on personal safety, of many types of freehold premises.

B – *Obligation to fence*

There is no implied obligation at common law on any owner of land to erect and maintain a fence or hedge around his land.[13] No

---

[11] *Halsall v Brizell* [1957] Ch 169 as explained in *Thamesmead Town Ltd v Allotey* [1998] 3 EGLR 97, CA, where it was said that taking a benefit of some kind did not suffice for the mutual benefit and burden principle to operate: there had to be a correlation between a burden and a benefit that the successor had to take. In the *Thamesmead* case, the succcessor was conferred by his conveyance no right to make use of landscaped and communal areas, having exercised a right to buy over a council house: he could not be required to pay a share of the maintenance costs as a result – he did not have the opportunity to choose, even in theory, to escape the burden of charges by not making use of the facilities in question.

[12] FR Crane 21 (1957) Conv (NS) 160, citing *Re Ellenborough Park* [1956] Ch 131, CA.

[13] *Hilton v Ankesson* (1872) 27 LT 519.

contract to fence in land may be inferred from the fact that an owner carried out occasional repairs.[14] A covenant by a freeholder to fence in land is purely personal and imposes no obligation on the covenantor's successors in title. Statute provides exceptions to this particular rule.[15]

C – Easement of fencing

An obligation by an owner to fence in the land against cattle is a "spurious easement" even though, in the absence of this right, an owner whose land is invaded by cattle may complain of cattle-trespass. If cattle escape from the servient land where the right is proved to exist, causing damage to the land or to its produce, the servient owner will be liable in damages.[16] This easement may be established by proof of a local custom, as well as by enclosure, but it does not require the servient owner to repair fences on the dominant owner's land.[17] The "fence" (whether wire or a hedge) must have been maintained and repaired by the owner or occupier, as a matter of obligation, for example at the demand of or following notice from the dominant owner, so as to render the obligation ancillary to the easement.[18]

This easement may be established by prescription at common law through long user as of right since time immemorial but subject to a presumption of lawful origin after 20 years' use. It has been asserted, *obiter*, that this easement may be acquired under the doctrine of lost modern grant.[19] This may be incorrect, since the easement cannot arise by express grant, owing to the rule that positive obligations cannot burden a successor in title to freehold land, and lost modern grant is based on a fiction of a lost but lawful grant.[20] The fencing easement has, however, been held to "lie in grant" and thus to be capable of implied acquisition as a result of section 62(1) of the Law of Property Act 1925.[21]

---

[14] *Boyle* v *Tamlyn* (1827) 6 B&C 329.
[15] Highways Act 1980 s 165(1); Railways Clauses Consolidation Act 1845 s 68.
[16] *Park* v *J Jobson & Son* [1945] 1 All ER 222, CA.
[17] *Egerton* v *Harding* [1974] 3 All ER 689, CA.
[18] *Jones* v *Price* [1965] 2 QB 618, CA.
[19] *Egerton* v *Harding, supra* at 691D (Scarman LJ).
[20] *Bryant* v *Foot* (1867) LR 2 QB 161.
[21] *Crow* v *Wood* [1971] 1 QB 77, CA; Crane (1971) 35 Conv (NS) 54.

II – Compensation for business tenants' improvements

A business tenant may obtain compensation on quitting for improvements to the demised premises, under Part I of the Landlord and Tenant Act 1927. In this way the landlord does not unfairly gain the full benefit from the work. The procedure, it is said, is of little present significance, but it is sometimes used by tenants as a way of obtaining the landlord's authorisation for improvements which are prohibited by the lease.[22] The following is an outline of this scheme,[23] whose abolition was proposed by the Law Commission as long ago as 1989.[24]

The 1927 Act applies to property used wholly or partly for the purposes of a trade, business or profession. Any improvement, whether by the tenant or his predecessor in title, qualifies for compensation. Trade or other tenants' removable fixtures do not qualify. Certain improvements, notably in pursuance of a statutory obligation, before October 1 1954, or under an obligation in the lease, do not qualify (s 2(1)). The Act is excluded if the tenant is entitled to remain on the premises at the expiry of his lease, as where he holds over under a continuation or new tenancy under Part II of the Landlord and Tenant Act 1954, and does not, in particular, apply to an improvement made less than three years before the termination of the tenancy (s 2(1)(c)).

A procedure, leading to a certificate from the landlord or, in default, the court, is obligatory[25] before compensation may be claimed. The tenant must serve notice of his intention to claim compensation with a plan and specification of the work (s 3(1)). He must require the landlord, once the work is done, to certify that it has been duly completed and in default, the tenant may apply to the court for a certificate (s 3(6)). He must then within strict time-limits claim compensation. These limits cannot be extended by the court.[26] The landlord may object, by notice served within three months of the tenant's notice of claim, to the improvement (s 3(1)). Failing an objection, the tenant may proceed, provided he does so

---

[22] Luxton and Wilkie, *Commercial Leases*, (1998) pp 167–168.
[23] For more detailed coverage see *Fox-Andrews on Business Tenancies*, Chapter 10.
[24] Law Com No 178.
[25] *Hogarth Health Club Ltd v Westbourne Investments Ltd* [1990] 1 EGLR 89, CA.
[26] *Donegal Tweed Co v Stephenson* (1929) 98 LJKB 657.

within the plan and specification submitted to the landlord. If the landlord objects, the tenant may apply to the court (usually the county court) which may certify the improvement to be proper (s 3(6)). In any case, the improvement, to qualify, must add to the letting value of the holding at the termination of the tenancy. It must be reasonable and suitable to the character of the premises. Nor may it diminish the value of any other property of the same or any relevant superior landlord (s 3(1)). The landlord may avoid paying compensation by carrying out the work in return for a reasonable rent increase (s 3(1)), provided he has offered to carry out the improvement.

Any compensation payable must not exceed either:

(a) the net addition to the value of the holding as a whole which results directly from the improvement; or
(b) the reasonable cost of carrying out the improvement at the termination of the tenancy, subject to a deduction of the cost of putting the improvement into a reasonable state of repair, so far as this cost is not covered by the tenant's repairing obligations (s 1(1) and 1(2)).

If the landlord intends to demolish or to structurally alter the premises or to change the use of the premises or any part after the termination of the tenancy, he may defeat the claim to compensation entirely or reduce it (s 1(2)). If the landlord has given the tenant or his predecessors in title any benefit in consideration for the improvement, this is taken into account in reduction of compensation (s 2(3)).

III – Rent review implications

Rent review clauses are many and varied. They generally have in common the aim of protecting the value of the landlord's rental income against inflation, by envisaging an increase in the rent level to that of the open market at regular, short intervals, which purpose has been recognised judically.[27] Rent review clauses are construed purposefully,[28] and artificial assumptions not expressly and clearly

---

[27] *Basingstoke and Deane Borough Council v Host Group Ltd* [1987] 2 EGLR 147 at 149, CA.
[28] *City Offices plc v Bryanston Insurance Co Ltd* [1993] 1 EGLR 126 at 128A.

set out in the clause will not ordinarily be made, since there is a presumption of reality in these matters.[29]

If, therefore, a rent review clause is silent as to the state of repair or condition to be assumed of the demised premises at the date of review, the premises are to be valued in their then state and condition,[30] unless the rest of the lease requires the valuer to value the premises in some other state of repair.[31] In the absence of an express provision allowing him to do so, a valuer cannot determine a variable rent depending on the state of repair of the demised premises.[32] Yet if a rent review clause requires a "reasonable" rent to be ascertained for the demised premises, this might indicate that any default in compliance with his repairing covenant by the tenant must be left out of account in assessing a reviewed rent.[33] Where a rent review clause provided that the tenant was obliged to perform or observe covenants, any diminishing effect on the rental value of the premises because the premises had been left out of repair by the tenant in breach of covenant was left out of account in assessing a reviewed open market rent. The court invoked the principle that a person cannot take advantage of his own wrong.[34]

If the tenant is not in breach of his repairing covenant, or if the work could be carried out, assuming his covenant had been broken, by a different method, then it is open to the tenant to alleviate or eliminate the effects of this particular principle. Hence, where a valuer assumed that the tenant had been in breach of covenant in not stripping out asbestos from the premises, the case was remitted for reconsideration since it was not established that such work was the only practical way of complying with the tenant's covenant. His failure to do the work concerned did not *ipso facto* constitute a breach of covenant.[35]

---

[29] *Co-operative Wholesale Society* v *National Westminster Bank plc* [1995] 1 EGLR 97.
[30] *Laura Investment Co Ltd* v *Havering London Borough Council* [1992] 1 EGLR 155.
[31] Bernstein and Reynolds, *Handbook of Rent Review*, para 4–52.
[32] *Clarke* v *Findon Developments Ltd* [1984] 1 EGLR 129.
[33] Cf *Dickinson* v *Enfield London Borough Council* [1996] 2 EGLR 88 (improvements).
[34] *Harmsworth Pension Funds Trustees Ltd* v *Charringtons Industrial Holdings Ltd* [1985] 1 EGLR 97. A specific version of the same deeming rule is that of the RICS/Law Society Model Clause, which assumes that the tenant has complied with his repairing covenants.
[35] *Secretary of State for the Environment* v *Euston Centre Investments (No 2)* [1995] CLY 3062.

Some rent review clauses state in terms that it is to be assumed for rent review purposes that the landlord has complied with his covenant to repair or to keep in repair and where relevant, that he has provided services. Such an assumption has the unfair effect that the tenant is assumed to be benefitting from premises in good condition and for services he is obtaining, both of which may be contrary to the fact, so that he is at risk of paying an artificially inflated rent after review,[36] arguably mitigated by the fact that his ability to obtain remedies such as specific performance of the landlord's covenants could be a relevant factor in the rent review process.[37] This factor cannot be decisive, since any equitable remedy, such as specific performance, depends for its availability on the discretion of the court.

As to the position where a rent review clause requires that the premises are fit for immediate occupation and use by the tenant, and are in a good state of repair and condition, the intention seems to be to prevent the tenant from claiming that he may obtain a discount at rent review on account of the actual state of repair of the premises at the review date. The Court of Appeal refused further to assume that a hypothetical tenant would have carried out further or different works from those in fact done by the actual tenant, so leading to a reduction in the open market rent he would be willing to pay at review.[38] This decision is a good example of the presumption of reality: if parties wish a clause to go further than required by reality and the purpose of rent review clauses, they must clearly say so.[39]

Trade fixtures, even where these have been installed under an obligation in the lease, which the tenant is entitled to remove at the end of the lease are not taken into account, in the absence of clear express contrary language, in assessing the value of the premises at a rent review.[40] Equally, equipment not fixed to the premises was not to be "rentalised" by a rent review assumption that the premises were to be fit for immediate use and occupation.[41]

---

[36] Bernstein and Reynolds, para 4–56.
[37] Bernstein and Reynolds, para 4–56.
[38] *London & Leeds Estates Ltd* v *Paribas Ltd* [1993] 2 ELGR 149.
[39] See also eg *Pontsarn Investments Ltd* v *Kansallis-Osake-Pankki* [1992] 1 EGLR 148.
[40] *Young* v *Dalgety plc* [1987] 1 EGLR 116; also *New Zealand Government Property Corpn* v *HM&S Ltd* [1982] QB 1145.
[41] *Ocean Accident & Guarantee Corp* v *Next plc* [1996] 2 EGLR 84.

The presumption of reality was held to require a valuer not to value certain premises for rent review purposes on the basis that the landlord, by the review date, would have carried out major works of improvement[42] so as to increase his rental income – as already seen, the tenant is paying for that which he occupies at the review date.[43] On the other hand, where a tenant had in fact carried out, under a pre-lease licence, certain improvements, the effect of these on the rental value of the premises had to be taken into account and he could not distinguish between the lease and the licence.[44] As with any of the principles here discussed, the canons of construction just mentioned would yield to the express language of a rent review clause, if clear enough to displace them. It would thus, given the variety of wording sometimes seen in rent review clauses, be unsafe to assume that the cases here discussed were any more than illustrations on particular facts of the application of the general principles of construction of rent review clauses.

IV – Effect of listed building controls

Two specific listed building controls within the planning system are here noted, the first of which affects alterations to listed buildings and the second of which governs those buildings which fall into disrepair.[45]

A – Alterations to listed buildings

Where a builing of special architectural or historical interest has been listed for planning purposes under the Planning (Listed Buildings and Conservation Areas) Act 1990, the carrying out without listed building consent of certain works to a listed building

---

[42] To carry out means either physically carrying out the works or arranging, contractually or otherwise, for a third party to do them, with involvement in the form of supervision and the like: *Durley House Ltd* v *Cadogan* [2000] 1 WLR 246.
[43] *Iceland Frozen Foods plc* v *Starlight Investments Ltd* [1992] 1 EGLR 126, CA.
[44] *Ivory Gate Ltd* v *Capital City Leisure Ltd* [1993] EGCS 76.
[45] For a detailed examination of the position, see Mynors, *Listed Buildings, Conservation Areas and Monuments* 3rd edn (1999) Chapter 9. For a criticism of the present rules, see Mynors "Do We Need Listed Building Consent?" [1998] JPL 101.

constitutes an offence. Those works are "any works for the demolition of a listed building or for its alteration or extension in any manner which would affect its character" as a listed building. The controls are wide: the external repainting of a listed building may fall foul of section 7 of the 1990 Act.[46]

The offence is constituted by section 9 of the Act: it is of strict liability which means that a criminal intention does not have to be proved against an accused.[47] The owner (or other person) who has carried out the works complained of has a statutory defence, the onus of proving which falls on him, if he is able to establish the following matters (s 9(3)):

(a) that the works to the building were urgently necessary in the interests of safety or health or for the preservation of the building;
(b) that it was not practicable to secure the above objects by works of repair or works for affording temporary support or shelter;
(c) that the works were limited to the minimum measures immediately necessary;
(d) that notice in writing justifying in detail the carrying out of the works was given to the local planning authority as soon as reasonably practicable.

The penalties laid down (s 9(4)) are a fine of up to £20,000 on summary conviction or for a conviction on indictment, a term of up to two years or a fine or both. Any financial benefit to the accused of the work concerned may expressly be taken into account.[48]

A local planning authority is empowered under section 38 of the 1990 Act to issue a listed building enforcement notice if it appears to them that any works are being or have been executed to a listed

---

[46] *Windsor and Maidenhead Royal Borough Council* v *Secretary of State for the Environment* [1988] 2 PLR 17, even though such work would be allowed by the General Permitted Development Order 1995, SI 1995 No 418, Sched 2, Part II, Class C.
[47] See *R* v *Sandhu (Major)* [1997] Crim LR 288 (where evidence which would be relevant to mens rea was accordingly held to have been wrongly admitted and to have rendered a conviction unsafe).
[48] See *R* v *McCarthy & Stone (Developments) Ltd* [1998] CLY 4198, where the maximum fine was imposed on a company one of whose directors, aware of listing, had ordered the demolition of the property in issue.

building in their area in contravention of section 9. The notice may be issued if the authority consider it expedient to do so having regard to the effect of the works on the character of the building – because it is of special architectural or historical interest (s 38(1)). Not later than 28 days after it is issued and not later than 28 days before it is due to take effect, a copy of the notice must be served on any owner, and any occupier of the building and on any person having an interest in the building who, in the authority's opinion, is materially affected by the notice (s 38(4)) – such as perhaps an owner's mortgagee.

The notice[49] must specify the alleged contravention and must require specified steps[50] to be taken:

(a) for restoring the building to its former state;
(b) if the authority consider that such restoration would not be reasonably practicable or would be undesirable, for executing such further works as specified in the notice as they deem necessary to alleviate the effects of the work carried out without listed building consent;
(c) for bringing the building to the state in which it would have been if the terms and conditions of any listed building consent had been granted.

If an owner of a listed building carries out repairs to it, if these affect or change the appearance of the builing, when compared to the materials originally used, it seems that a listed building enforcement notice could properly be served on the owner, requiring him to remove the offending materials and to carry out the repairs within a specified time-scale with the materials previously used. Otherwise, a cost-cutting but irresponsible owner could carry out cheap work of repair to a roof or other part of the building and still avoid prosecution on the ground that no major works had been carried out. Thus, an owner who replaced natural roof slates with asbestos slates was required by the High Court, in

---

[49] Against which the owner etc may appeal to the Secretary of State under s 39. However, there may be insufficient safeguards in this procedure to satisfy art 6(1) of the European Human Rights Convention: see The Times, December 14 2000.
[50] Within a period or periods – applying for different steps – specified in the notice (s 38(3)).

an amended notice, to remove the new asbestos slates and to carry out the roofing work with natural slates.[51] Equally, so preventing the Act being side-stepped, where a listed building had been demolished so that some 80% of the timbers had been preserved, with a view to their shipment to the USA, and some foundations and footings remained, the High Court upheld a notice which required the restoration, correctly so labelled, of the building, using the timbers which had come into the safe custody of the local authority following earlier proceedings.[52] As noted in this case, however, there may come a point where restoration is not possible, as where the building is reduced to ashes and rubble which cannot be sensibly reconstituted. Once that happens, there appears to be nothing within these controls which the authority can do.

B – Repair and acquisition of listed buildings

A local authority is given a statutory default power (s 54 of the 1990 Act) to execute any works which appear to them to be urgently necessary for the preservation of a listed building in their area (s 54(1)). This power relates only to an unoccupied building or to any unoccupied parts of it (s 54(4)) and the owner of the building must be given a minimum seven days' notice, which describes the works to be carried out (s 54(5) and (6)). Provision is made for the recovery of the cost of the works from the owner (s 55), who has a right of appeal to the Secretary of State, so that, notably, he may contend that some or all of the works were unnecessary for the preservation of the building.

If the Secretary of State is of the opinion that reasonable steps are not being taken for the preservation of a listed building, as where it is in need of repair, he has a power to compulsorily acquire the building, or he may authorise the relevant district council or London Borough Council to acquire the building (s 47(1) of the 1990 Act). The power is not to be used unless, first, a repairs notice has been served under section 48 of the Act and it has not been withdrawn (which it may be at any time (s 48(3)) and two months have elapsed from the service of this notice. The House of Lords has limited the scope of this provision by ruling that "preservation" in

---

[51] *Bath City Council v Secretary of State for the Environment* [1983] JPL 737.
[52] *R v Leominster District Council, ex parte Antique Country Buildings* [1988] JPL 554.

section 48 refers only to works of repair reasonably necessary to preserve the building as at the date it was listed, as opposed to works of restoration or renewal. But, since a notice is a formality, a notice which specified works which went beyond repairs and were described as "restoration items" was not rendered invalid by the excessive specifications and so, on the owner not complying with the notice, the Minister had been entitled to confirm a complusory purchase order.[53]

V – Options to break, renew or purchase

Some leases contain a tenant's option to break, to renew or to purchase the freehold. The exercise of the tenant's right is commonly subject to an absolute condition precedent, to the effect that he has, at the operative date[54] paid all the rent due and performed all the covenants of the lease. In these circumstances, a breach of covenant to repair or to keep in repair which is subsisting, in the sense of giving the landlord a right of action against the tenant as at the operative date, will be fatal to the exercise by the tenant of his right to break, renew or purchase.[55] Although in the case of an option to terminate the lease it may be of greater importance to the landlord to see to it that the lease is "clear" and that there are no subsisting breaches of covenant than in the case of a tenant's option to renew or purchase, where the tenant will be staying on, there is no distinction of principle between the three classes of case.[56] It has, however, been suggested that where there is a requirement that the tenant has "reasonably" performed his obligations, a distinction may exist between options to renew and to break, since in the latter case strict compliance by the tenant, with, say, a covenant to redecorate is more important than in the former, where he is staying on.[57]

By "spent" is meant a breach in respect of which the landlord has no subsisting cause of action at the operative date of the tenant's

---

[53] *Robbins* v *Secretary of State for the Environment* [1989] 1 WLR 201.
[54] This may be eg the date of exerise of the right, that of the termination of the lease or of expiry of the tenant's notice.
[55] *Bass Holdings Ltd* v *Morton Music Ltd* [1987] 2 EGLR 214, CA.
[56] According to Kerr LJ in *Bass Holdings Ltd* v *Morton Music Ltd, supra* at 216A–B.
[57] See *Reed Personnel Services Ltd* v *American Express Ltd* [1997] 1 EGLR 229 at 230 (Jacob J).

rights. It is not an implied condition precedent of a right to renew, break or purchase that the tenant has never committed any past breaches of covenant – such a requirement would be virtually impossible to insist on and would render the rights under the option worthless. Thus, where a tenant had in the past broken his covenants and the landlord forfeited the lease, but relief was granted to the tenant, those events, being spent, did not affect the tenant's right to renew.[58]

If a breach is subsisting, its effect is to destroy the tenant's right to renew, to break or to purchase as the case may be, even if it is not substantial in nature. Exact compliance with the covenant at the operative date is required.[59] Thus, in one case the tenant failed to comply with a condition precedent that the premises must be redecorated in what would have been the last year of the term if the lease had succesfully been determined. In fact, they had been repainted just before then: nevertheless, the covenant had not been complied with and the option to renew could not be exercised.[60] The court refused to rewrite the lease so that the right to renew survived the breach even though no substantial damages could be recovered for the breach from the tenant.

Some mitigation of the severity of these principles may be afforded by the fact that in the case of a tenants' covenant to repair, the covenant is broken by a failure to repair within a reasonable time. Thus it has been said, that if there is a disrepair which has occurred at the last minute and which "cannot be remedied as the clock strikes 12", this would not deprive the tenant of his rights under an option clause.[61] It remains to be seen whether, if this notion is followed, it would also apply to tenants' covenants to keep in repair, where a breach arises as soon as a fact of disrepair comes into existence. In such cases, a last-minute breach could perhaps be disregarded as being technical or trivial under the *de minimis* principle. At all events, a tenant's option could be exercised

---

[58] *Bass Holdings Ltd* v *Morton Music Ltd, supra.*
[59] *Dun & Bradstreet Software Services (England) Ltd* v *Provident Mutual Life Ass..rance Association* [1998] 2 EGLR 175 at 180B (Peter Gibson LJ), following *Finch* v *Underwood* (1876) 2 ChD 310 (subsisting repairs costing a small sum to remedy).
[60] *Bairstow Eves (Securities) Ltd* v *Ripley* [1992] 1 EGLR 47.
[61] *Trane (UK) Ltd* v *Provident Mutual Life Assurance* [1995] 1 EGLR 33 at 37E.

where the form of the condition precedent was to require only reasonable compliance with the tenant's covenants, which had been complied with sufficiently.[62]

VI – Liability of tenant in respect of fixtures

A – Introduction

An express tenants' covenant to repair or to keep in repair extends to any chattel affixed to the land or any building on the land, comprised within the demised premises, which has become a fixture such that ownership of it passes with the landlord's reversion.[63] The House of Lords[64] has approved a revised test to determine whether a chattel brought onto land has become a fixture and so part of the land. While the classical formulation involves two tests long used by the common law, the degree of annexation and the purpose of annexation, the revised test identifies three classes of case: (a) chattels; (b) fixtures; and (c) items which become part and parcel of the land itself. Objects in (b) and (c) are treated as being part of the land. Since fixtures have traditionally been considered to be chattels attached permanently to the land or to a building on land so as to become part of the land, the court asking whether a chattel could be removed without injury to itself or to the land, it is difficult at first sight to see what this new formula adds to the classical tests, other than to confirm the result of their joint operation.[65] Notwithstanding the new test, uncertainties are likely to remain, as it is not always an easy matter to apply the law to given facts.[66]

B – Revised test to decide if item is a fixture

In reformulating the test governing whether a chattel had become a fixture, their Lordships were concerned with the issue of whether

---

[62] *Gardner* v *Blaxill* [1960] 1 WLR 752.
[63] See HW Wilkinson "Chattels, Fixtures and Land" (1997) 147 New Law J 1031.
[64] In *Elitestone Ltd* v *Morris* [1997] 2 EGLR 115. See Conway [1998] Conv 418.
[65] This seems to have been the process of reasoning in *Chelsea Yacht & Boat Co Ltd* v *Pope* [2000] 22 EG 147.
[66] See eg *Botham* v *TSB Bank plc* (1997) 73 P&CR D1; [1996] EGCS 149; Haley [1998] Conv 137.

a chalet or bungalow on a lot, which rested on concrete foundation blocks in the ground, was a fixture or a chattel. The item could not be detached from its supports without destroying it. The House of Lords unanimously held that the chalet or bungalow had become part of the land. It had been built in such a way that it could not, without destroying it, be removed from the land, to which it had become annexed permanently. The House of Lords said that each case must be examined individually on the facts. Lord Steyn referred both to the traditional test of degree of annexation and purpose of annexation. These accordingly remain relevant tests. Thus, Lord Steyn referred to an Australian case,[67] when discussing the degree of annexation, under which the courts have looked at the extent to which the chattel is permanently attached to the land. In this case, a wooden house erected on a town allotment, and not fixed permanently into the soil, but resting on its own weight on brick piers, was held a fixture. However, although in the case of a large, heavy, object such as a house[68] "annexation goes without saying",[69] questions of greater factual difficulty may arise in intermediate cases[70] and especially with smaller objects.[71] At that point the objective intentions of the person bringing the item onto the land become relevant, as derived from the test of degree and object of the annexation.[72] As was pointed out by Blackburn J:[73]

---

[67] *Reid* v *Smith* (1905) 3 CLR 656; also now *Chelsea Yacht & Boat Co Ltd* v *Pope, supra*.
[68] Or sculptures not fixed into the building, but part of the design of a hall, they were still held fixtures (*D'Eyncourt* v *Gregory* (1866) LR 3 Eq 382) as were seats fastened to the floor of a cinema by screws (*Vaudeville Electric Cinema* v *Muriset* [1923] 2 Ch 74).
[69] Per Lord Steyn in *Elitestone Ltd* v *Morris, supra* at 117A.
[70] Thus in *Pan Australian Credits (SA) Pty Ltd* v *Kolim Pty Ltd* (1981) 27 SASR 353, air-conditioning units installed in the roof of a convention centre, fixed to the roof by bolts and connected to electric cables in ducts in the ceiling of the premises were fixtures.
[71] So that eg a large statue resting on its own weight on a plinth was not part of the realty (*Berkeley* v *Poulett* [1977] 1 EGLR 86, CA) any more than were light fittings and fitted carpets in a flat (*Botham* v *TSB Bank, supra*). Likewise, easily removable office partitions were not fixtures (*Short* v *Kirkpatrick* [1982] 1 NZLR 358) any more than was wood pannelling (*Spyer* v *Phillipson* [1931] 2 Ch 183).
[72] *Elitestone Ltd* v *Morris, supra* at 117H (Lord Steyn).
[73] In *Holland* v *Hodgson* (1872) LR 7 CP 328 at 335.

Blocks of stone placed on the top of another without any mortar or cement for the purpose of forming a dry stone wall would become part of the land, though the same stones, if deposited in a builder's yard and for convenience sake stacked on top of each other in the form of a wall, would remain chattels.

In other words, if the intention of the person attaching the chattel to the demised premises is to enhance their value and enjoyment, the item, even if resting by its own weight on the land, may be a fixture and considered as part of the land. On the other hand, if an item is easy to detach from the land or premises without causing lasting damage, and it was not installed with any idea of permamence but rather as a moveable item, as with a greenhouse which could be dismantled and removed without permanently injuring the land,[74] it is more likely that it will be held to have retained its original nature as a chattel. Sometimes difficulty has been caused where an item, apparently permanently attached to land, is removable at the end of the lease by the tenant. Thus a 135-foot-long and 50-foot-wide shed built on a concrete floor to which it had been attached with iron straps was held to have become part of the land, and yet to be removable by the tenant on expiry of his lease as a tenants' trade fixture.[75]

Once an item, such as central heating equipment or air-conditioning plant, or a heating unit, is a landlords' fixture, it will fall within a tenant's covenant to repair, unless expressly excluded from it, provided it is situated within the demised premises.[76] In the interests of certainty, the tenant may wish to have complied a list of what items are to be considered to be landlords' fixtures prior to taking his lease. If landlords' fixtures such as central heating systems or the air-conditioning equipment fail during the term of the lease, the tenant may find himself having to pay for their replacement with the nearest modern equivalent, which in turn may render it advisable for him to satisfy himself, prior to the lease being taken, that such items are in proper working order.[77]

---

[74] *Deen* v *Andrews* [1986] 1 EGLR 262 (large but removable greenhouse worth over £2,000).
[75] *Webb* v *Frank Bevis Ltd* [1940] 1 All ER 247, CA.
[76] See *Shortlands Investments Ltd* v *Cargill plc* [1995] 1 EGLR 51 (so that a heating unit which was a landlords' fixture could not be removed by the tenant without replacing it, as the tenant had covenanted to yield up the premises and fixtures in good tenantable repair).
[77] Ross, 4th edn, para 3.20.

VII – Human rights

On October 2 2000, European Convention on Human Rights became part of English law.[78] This book is not the place for any detailed examination of that constitutional innovation, the Human Rights Act 1998. The courts are, in particular, enjoined by it to read and give effect to primary and delegated legislation, so far as possible, in a way which is compatible with Convention rights (s 3(1)). If primary legislation is incompatible with Convention rights, it continues to be valid, operational and enforceable (s 3(2)) but the High Court or above may declare that it is incompatible with the Convention (s 4).[79] Human rights challenges in dilapidations may arise under article 1 of the First Protocol to the Convention, which refers to the entitlement of any person to the peaceful enjoyment of his possessions.[80] The High Court has pointed out that the Act makes no reference to contracts entered into prior to its commencement,[81] and it may be, not least owing to the principle of non-retrospectivity of legislation, these will continue to be interpreted in accordance with established common law principles. However, in future it will be necessary for a court to disregard a precedent whose effect would be to interpret legislation or contractual terms in a way which is inconsistent with Convention rights. Only time and litigation will disclose the impact of possible human rights challenges to the law of dilapidations. It is to be hoped that these will not destabilise an area of law where until now certainty and predictability have carried a premium.[82]

---

[78] Human Rights Act 1998 (Commencement No 2) Order 2000, SI 2000 No 1851. See Harpum "Property Law – The Human Rights Dimension" (2000) 4 L&T Rev 4 and 29.
[79] Incompatible delegated legislation may in principle be quashed (s 3(2)).
[80] As noted in Chapter 7, peaceable re-entry might be vulnerable to challenge. As to the effect of article 6 of the First Protocol on residential tenancy repossession procedures, see Henderson 144 (2000) Sol Jo 906.
[81] *Biggin Hill Airport Ltd* v *Bromley London Borough Council* The Times, January 9 2001.
[82] See *Phillips* v *Mobil Oil Co* [1989] 2 EGLR 246 especially at 249.

Index

Access to Neighbouring Land
    access orders under.. 258–262
    basic preservation works, under 260–261, 262–264
    common law, right to 257
    defences of respondent to order for 261
Agricultural Dilapidations
    farm business tenancies, under 230–231, 241
    general deterioration damages........................... 239–240
    model clauses, rules as to................................. 232–235
    particular parts of holding damages...................... 236–239
Alterations to Premises
    absolute prohibition against............................... 90–91
    improvements, as 94–96
    qualified prohibition against 94–96
    statute, effect on 91–94
Ameliorating Waste
    meaning of... 18–19

Building Lease
    covenant to erect buildings in 89–90
    damages in case of 105–106
    scope of covenant to repair in 88–89
Buildings
    additional... 88–90
    agricultural holdings, in relation to 232–236, 236–237, 238
    constituting nuisance 242–248
    constituting statutory nuisance 206–211
    dangerous 228–229
    insurance by landlord, of 171–178
    party walls between 212–227
Business Tenancies
    improvements compensation............................. 269–270
    rent, effect of disrepair in relation to.......................... 201
    repairs under 201–202
    service charges 38–39, 43–45, 157–164

Commonhold Tenure
  schemes as to... 13–14
Compensation
  deterioration of agricultural holdings..................... 236–240
  improvements, for.. 269–270
Contaminated Land
  clean-up... 55–56
Contract
  collateral... 48–49
  liability to visitors entering under....................... 249–255
Covenant to Repair
  dependent.. 50–51
  express... 51–53, 66–75
  forfeiture for breach of................................. 117–143
  implied.. 5–7, 26–28, 31–34
  improvements distinct from........................... 54–55, 71–73
  independent.. 50–51
  inherent defects within................................... 73–75
  interpretation of.................................... 7–8, 66–75
  notice under... 56–58
  rebuilding outside....................................... 53–54
  standards under.. 86–88
  statute, under................................. 179–201, 202–206
  subordinate renewal within............................... 77–80
Covenants
  alterations to premises.................................. 90–96
  default.. 101–102
  general enforcement of................................... 63–64
  insurance.. 171–178
  repair (*see* Covenant to Repair *and* Repair)

Damages
  landlords'claims:
    during lease.................................... 104–106, 110
    end of lease............................... 103–104, 106–110
    tenants'claims...................................... 145–148
  special classes of award:
    agricultural tenancies.............................. 236–241
    protected tenancies..................................... 111
    requisitioning.. 111
    subleases... 112
  waste claims... 20
Dampness
  causes of... 3–5
  remedies for.......................... 71, 145–148, 148–153

Index

Dangerous Buildings or Structures
    control of 228–229
Decorative Repair
    relief for 85
    required by repairing obligation 84–86
Defective Premises
    duty of builder or landlord 31
    implied covenants in respect of 26–28, 28–31, 191–194
Definitions
    demised premises 45
    installations 185–186, 186–187
    structure and exterior 45, 184–185
    windows 45–46
Disability Discrimination
    alterations to avoid 92–93
Disrepair
    fitness for use, contrasted with 1–2, 3–5, 11–12, 23–24, 28–30,
    .... 31–32, 34–36, 47–48, 71, 179–180
    proof of 35, 70–71
    rent review, effect of 96–97, 270–273
    statistics as to 2–3
    tenants' options, effect on 277–279

Entry to Repair
    express right 58–59, 242–243
    implied right 59–60, 243
    statutory rights of 58, 189–190, 221
Equitable Remedies
    set-off 148–151
    specific performance:
        award to landlord 114–116
        award to tenant 151–153
Extended Lease
    repairing liability under:
        extended lease 200
        leasehold enfranchisement 200–201

Fair Wear and Tear
    exception for 83–84
Fence
    obligations to 267–268
Fire (*and see* Insurance)
    insurance against 171–175
    reinstatement after 175–178

Fitness for Habitation
  implied warranties as to. 28–32
  reforms as to. 11–12, 34–36
Fixed Equipment
  liability to maintain. 232–235
Fixtures
  definition of 279–281
  removal of, as waste. 15, 17
Flats
  common parts of 27–28
  demise, extent of 27–28
  extended obligations of landlord. 188–189
  manager of. 155–156
  receiver of rents of 154
  service charges, rules as to:
    common law 157–164
    statutory. 164–169
Forfeiture
  decorative disrepairs. 136–137
  effect of statute on 117–118, 124–141
  insolvency, leave requirements for 120
  landlords' costs in. 135–136
  peaceable re-entry as. 120–121
  position of mortgagees 132–135
  relief against. 129–132, 136–137
  reform of law of. 118–120
  schedules of dilapidations in 127–128
  statutory restrictions on:
    general 124–129
    specific to repairs. 137–143
    status of lease during 123–124
  waiver of right to 121–123
Freeholder
  fencing rules respecting 267–268
  nuisance liability of. 242–247
  obligation to carry out repairs of. 265–267
  occupiers' liability of. 249–257
  repair notices against 202–206
  statutory nuisance proceedings against 206–211
French Law
  flats, as to. 14
  implied repairing duties of landlord. 41–43
Furnished Houses
  fitness, warranties of landlord, as to 31–32
  obligations, landlord, to repair 26–27

Index

General Enforcement of Covenants
  fixed charges 102–103
  reforms of law as to 63–64

Highway
  liability in nuisance for premises abutting 244–246

Human Rights
  dilapidations and. 121, 275, 282

Implied Obligations to Repair
  common user parts 27–28
  French law as to 41–43
  furnished houses. 31–32
  general principles as to. 23–26
  reform of law as to 8–11, 34–36, 39–40, 180–181
  Scottish law as to. 10–11
  statutory. 181–191

Improvements
  alterations as 94–96
  compensation for. 269–270
  covenants against 90–96
  disregard of 96–98
  repairs, distinguished from 37–38, 54–55, 64–65, 71–73, 187

Inherent Defects
  liability to remedy 3–5, 46–47, 55, 71–73

Installations
  statutory covenant to repair. 185–187

Insurance
  landlord, by. 171–178
  tenant, by. 171–178

Landlord
  adjoining owners, liability to 242–247
  alterations, consent of to. 94–96
  collateral warranties by 48–49
  collection of service charges by. 38–389, 43–45, 157–170
  contaminated land, liability of. 55–56
  covenant to insure by 171–178
  damages claims by:
    end of lease. 103–105, 106–110
    during lease 103–106
  defective premises, duties of 191–193
  entry to repair by. 58–60, 189–190
  express repairing obligations of. 37–60
  fitness, warranties by 46–48

forfeiture proceedings by 117–143
  implied obligations to repair of............................26–28
  manager's appointment against 155–156
  notice of lack of repair, to..................................56–58
  nuisance liability of......................................242–247
  receiver's appointment against...............................154
  reinstatement by 175–178
  set-off of rent against 148–151
  specific performance in relation to..................114–116, 151–153
  special cases, obligations of 194–202
  structural repairs, liability for 46, 184–189
  third parties, liability to 242–247
  visitors, liability to 249–257
  warranties of fitness by 28–32
Law Reform
  express repairing obligations 8–11, 39–40
  forfeiture..118–120
  implied repairing obligations..............................34–36
  law of waste 21
  specific performance...153
Lease
  agricultural 230–241
  business 201–202, 269–270
  building 89–90
  forfeiture of 117–143
  long...88, 194, 200–201
  short 6–7, 11–12, 32–34, 87, 145, 181–191, 195–199
Listed Building Controls
  alterations, limited by....................................273–275
  repairs, effect on 275–276
Local Authority
  dangerous structures control by 228–229
  repair notices served by 202–206
  statutory nuisance powers of..............................206–211
Long Leases
  alterations under...90–96
  lessee's liability to repair under.....................64–75, 77–86
  lessor's liability under................................37–39, 46–56
  service charges under......................43–45, 157–164, 164–170
  standard of repair under 88

Mortgage
  relief against forfeiture 132–135

Index

Notice
  abatement 207–208, 210–211
  damages claim, landlord's, prior to. 99–100
  repair, on 101–102
  repair 203–205
  repairs, as condition precedent to executing 56–58
  statutory forfeiture. 124–129, 139–143
Nuisance
  buildings adjoining highway, liability for. 244–246
  landlord liability for 242–247
  licence to commit 243–244
  liability of lessee in 247–248
  statutory. 206–211

Occupiers' Liability
  common duty of care 250–254
  exclusion of duty. 254–255
  occupier, meaning of. 249–250
  trespassers, duty to 255–257
Options
  breach of covenant to repair, effect on. 277–279

Painting and Decorating Covenants
  agricultural tenants, by. 234–235
  general covenant, to carry out 84–86
  relief for breach of. 85
  separate covenant, to execute 84–86
  standard of 87
Party Walls
  background to rules as to 212–215
  definition of 215–217
  dispute resolution in relation to. 225–227
  effect of statute on existing rights over. 217
  existing party structures 217–220
  new party structures. 224–225
  rights of building owner over 220–224
  withdrawal of support of 214

Rebuilding (*and see* Renewal and Repairs)
  meaning of. 53–54
  repairs, contrasted with. 7–8, 69–70, 77–81
Receiver and Manager
  equitable jurisdiction to appoint 154
  statutory jurisdiction to appoint 155–156

Relief against Forfeiture
  decorative repairs, in respect of. 85
  peaceable re-entry, despite. 121
  right to 118, 121, 129–135
Remedies for Breach of Covenant to Repair
  agricultural holdings. 236–240
  entry to repair 101–103
  forfeiture. 117–143
  landlord's damages 100, 103–114
  receiver's appointment. 154–156
  repudiation of lease. 144–145
  tenants' damages. 145–148
  set-off 148–151
  specific performance 114–116, 151–153
Renewal
  rebuilding compared to 53–54, 77–80
  repairs contrasted with. 7–8, 52–53, 64–65, 66–68, 69–70, 77–83
Rent
  set-off against 148–151
Rent Review
  disrepair, effect on. 270–273
  improvements, disregard in. 96–98
Repair
  additional buildings 88–89
  after notice. 101–103
  covenant to keep in. 51–53, 76–77
  dependent covenant to. 50–51
  entry to 58–60, 101–102, 189–190
  fair wear and tear, exception to. 83–84
  fixed equipment, of. 232–235
  forfeiture for breach of covenant to. 117–143
  French law as to 41–43
  general interpretation of covenants to. 1–2, 7–8, 37–38, 51–54, 55,
    . . . . 61–63, 64–71, 73–75, 77–83
  implied covenants to:
    landlords'. 5–7, 27–28
    tenants'. 32–34
  improvements, distinct from 54–55, 71–73
  independent covenant to 50–51
  inherent defects, as 73–75
  keep in. 76–77
  landlords' obligation to 51–55
  leave in 77
  meaning of 77–83
  notice of 56–58

Index

notice to 101–103
put in 75–76
redecoration as 84–86
reform of rules as to. 8–12, 34–36, 39–40
renewal, as distinct from 53–54, 69–70, 77–81, 82–83
Scottish law as to. 10–11
standard of. 86–88
statutory liability to 1, 6, 26, 29, 179–194, 202–205
tenants' obligation to 61–90
Repair Notices
   generally 203–206
   serious disrepair 205
Repairs
   access to neighbouring land for. 257–264
   decorative 84–86
   entry to carry out 58–60, 101–102, 189–190
   improvements, distinct from 37–38, 54–55, 64–65, 71–73, 187–188
   landlords' obligations (*see* Landlord)
   notice to execute 101–103
   party walls, to 212–227
   standard of. 86–88
   subordinate renewal as 52–53, 69–70, 77–80, 82–83
   structural 46, 184–185, 186–187, 188–189
   tenants' obligations (*see* Tenant)
Residential Premises
   assured tenancies 196–197
   extended leases. 200
   lease-back. 200–201
   protected tenancies 195–196
   secure tenancies. 197–199
Rights of Access and Entry
   basic preservation work, for. 257–264
   execution of repairs under lease 58–59, 59–60, 189–190, 242–243
   party walls, in relation to. 221
   statutory rights of 58, 189–190, 221

Schedule of Dilapidations
   contents of 127–128
Scottish Law
   repairing obligations of landlord. 10–11
Set-off
   common law 148–149
   equitable. 150–151
Service Charges
   general controls 157–158, 160–163

interpretation of 43–45
overpayments of under mistake 158–159
statutory consultation rules as to......................... 164–168
trusts of 163–164
Specific Performance
  equity jurisdiction 114–116, 151–153
  statutory jurisdiction 153
Standard of Repairs
  long leases... 88
  short leases 87
Statutory Liability to Repair
  assured tenancies 196–197
  business tenancies....................................... 201–202
  defective premises....................................... 191–194
  extended long leases.................................... 200–201
  following repair notice 202–206
  long leases... 194
  Rent Act tenancies...................................... 195–196
  residential landlord's liabilities...... 1, 5–7, 11–12, 26–28, 29–30, 31–32,
  179–191
  secure tenancies.. 197–199
Statutory Notices
  abatement..................................... 207–208, 210–211
  dangerous structure 228–229
  forfeiture....................................... 124–129, 137–143
  party walls 218–220, 224–225
  repair 203–206
Statutory Nuisance
  abatement notices............................. 207–208, 210–211
  appeals 210–211
  defences to notice 209–210
  enforcement procedure................................. 210–211
  local authority, duties in respect of 207–208
Structure
  covenants against alterations to........................... 90–96
  meaning of................................... 46, 184–185, 188
  repairs to 46

Tenant
  additional buildings, repair by 88–90
  alterations by.. 90–96
  damages awards to..................................... 145–148
  decorative repairs by 84–86
  deductions from rent by................................ 148–149
  express repairing obligations of........................... 61–90

fair wear and tear exception for. 83–84
fixtures, repairing liability of 279–281
general obligation to repair, of. 61–90
implied obligation to repair, of 32–34
improvement of premises by 64–65, 71–73, 269–270
inherent defects, liability of 73–75
manager, application to appoint by. 155–156
nuisance liability of. 247–248
obligation to:
    keep in repair. 76–77
    leave in repair 77
    put in repair. 75–76
occupier, liability as 249–257
rebuilding, obligation to execute. 66–68, 69–70, 77–80, 82–83
receiver, appointment of. 154
redecoration by 84–86
reinstatement by landlord 175–178
relief against forfeiture for 129–132
service charges payable by 38–39, 43–45, 157–168
set-off by 150–151
specific performance against 114–116
specific performance for. 151–153
standard of repairs of 86–88
subordinate renewal by 69–70, 73–75, 77–80, 82–83
waste, liability of. 19–20
Third Parties, Liability
    landlords. 242–247, 249–257
    tenants. 247–248, 249–257
Trespasser
    liability of occupier to. 255–257

Unfitness for Human Habitation
    disrepair, as opposed to. 1–2, 3–5, 6–7, 179–181
    meaning of 203–204, 206
    reform of law of. 11–12
Unfurnished Flats
    implied obligation of landlord of 27
Unfurnished Houses
    implied obligation of landlord as to 26–27

Voluntary Waste
    meaning of. 17–18

Waiver
    right of forfeiture, of 121–123

Warranty of Fitness or Suitability
    express 46–48
    implied 23–26, 28–32
    statute-implied 29
Waste
    ameliorating 18–19
    definition 15–17
    extent of liability for 19–20
    permissive 18
    reform of law of 21
    remedies for 20
    voluntary 17–18